Connected, Autonomous, Shared, and Electric Vehicles in the New Age of Mobility

Jesús F. Lampón
https://orcid.org/0000-0001-9267-4448
University of Vigo, Spain

Hugo Pérez-Moure
https://orcid.org/0000-0002-2233-0057
University of Vigo, Spain

IGI Global
Scientific Publishing
Publishing Tomorrow's Research Today

Vice President of Editorial Melissa Wagner
Director of Acquisitions Mikaela Felty
Director of Book Development Jocelynn Hessler
Production Manager Mike Brehm
Cover Design Jose Rosado

Published in the United States of America by
IGI Global Scientific Publishing
701 East Chocolate Avenue
Hershey, PA, 17033, USA
Tel: 717-533-8845 | Fax: 717-533-7115
Website: https://www.igi-global.com E-mail: cust@igi-global.com

Library of Congress Cataloging-in-Publication Data

LCCN: 2025036609 (CIP Data Pending)
ISBN13: 9798337344959
Isbn13Softcover: 9798337344966
EISBN13: 9798337344973
British Cataloguing in Publication Data
A Cataloguing in Publication record for this book is available from the British Library.

Table of Contents

Detailed Table of Contents

This chapter explores the strategic transformation and sustainability imperatives shaping the evolution of Connected, Autonomous, Shared, and Electric (CASE) mobility. It examines how digitalization, decarbonization, and new business models are redefining automotive value chains and enabling a shift from product-centric to service-oriented ecosystems. Integrating theoretical perspectives such as dynamic capabilities, the resource-based view, and circular economy principles, the chapter analyzes how organizations leverage innovation, data, and collaboration to achieve sustainable competitiveness. Through global case studies of leading mobility pioneers, it highlights emerging opportunities, policy enablers, and challenges in realizing sustainable, resilient, and inclusive mobility ecosystems. Ultimately, it emphasizes the need for integrated strategies that align technological advancement with environmental stewardship and socio-economic value creation in the new age of mobility.

With environmental, social, and governance (ESG) standards at the center of the international business strategies, there is an ever-growing pressure on the automotive industry to be able to guarantee ethical procurement sound carbon footprints and supply chain transparency. This paper examines the lucrative way in which Artificial Intelligence (AI) and Blockchain are soon to transform the sphere of ESG compliance and become the source of end-to-end supply chain visibility. It explores existing issues, namely, fragmentation of data, the lack of its traceability, and the inappropriateness of regulations. The study states the importance of AI to forecast sustainability threats, and the ability of the Blockchain to deliver immutable records of sourcing, logistics, and emissions. It also assesses regulatory bodies, the scale-related problems, and proposes a road map towards incorporation of such technologies, and finally, gives a model of how digital change can be implemented in the creation of responsible, sustainable, and smart networks of automobiles.

Sandeep Gupta, Graphic Era University, Dehradun, India
Devendra Kumar Tiwari, United Institute of Management FUGS, Prayagraj, India
Chandra Shekhar, United Institute of Management FUGS, Prayagraj, India
Kiran Saroj, United Institute of Management FUGS, Prayagraj, India

This chapter discusses the role of sustainability certifications in driving the adoption of sustainable practices across various industries. It outlines the significance of these certifications in promoting responsible behaviour and their global impact on industrial transformation. The chapter categorizes different certification types, including environmental and social responsibility certifications, and presents case studies from sectors such as manufacturing, retail, construction, and tech. It examines the influence of certifications on supply chains, product development, and innovation while also addressing regulatory compliance and industry incentives. Furthermore, it anticipates future trends, emphasizing the role of digital tools and AI, and highlights the integration of certifications within the circular economy. The concluding perspective focuses on the sustainability landscape projected for 2030 and the vital importance of certifications in industrial sustainability.

Hugo Sánchez Nieto, Universidad de Almería, Spain
Nieves García de Frutos, Universidad de Almería, Spain

The study delves on the antecedents of sustainable vehicle adoption (electric and hybrid cars) in Spain. Employing the Comprehensive Action Determination Model (CADM) as a theoretical framework, it accounts for four types of psychosocial factors, such as intentional variables, normative influences, situational variables, and habitual behaviour. Additionally, sociodemographic factors and pro-environmental behaviours considered to obtain a more complete profile of the owner. After analysing the data from 358 individuals, the results demonstrate that the CADM model effectively explains the adoption of sustainable cars, indicating that intention and perceived behavioural control are the primary determinants of sustainable car adoption. Additionally, significant differences among (non)owners are observed across sociodemographic, psychographic, and behavioural factors. Implications and practical recommendations are provided, emphasising the necessity for policies and initiatives to address psychosocial barriers and facilitate transition towards environmentally sustainable mobility.

Congestion, pollutants, and infrastructure stress are some of the growing issues faced by the transportation industry. This chapter analyzes present uses, difficulties, and future directions to examine sustainable AI solutions as a transformational force. It highlights carbon reduction and ethical governance while examining AI's involvement in autonomous vehicles, traffic optimization, predictive maintenance, and upcoming systems like the Hyperloop through a thorough literature analysis. From 1995 to 2025, bibliometric studies show changing research patterns that emphasize multidisciplinary connections with Industry 4.0, healthcare, and climate action. Key findings highlight AI's ability to address obstacles including data quality, cybersecurity, and equity while aligning with UNSDGs for resilient, low-emission mobility. In order to promote smarter, greener transportation ecosystems, this chapter advocates for the ethical and scalable deployment of AI by filling gaps in systemic integration.

This study analyzes the impact of automation and digitalization on employment through the lens of Skill-Biased Technological Change (SBTC), using the Toyota Guanajuato plant (TMMGT) as a case study. Using a qualitative methodology, the study identified an organizational strategy centered on internal relocation, continuous training, and upward mobility. This has allowed for the integration of new technologies without generating labor disruptions. The findings show that the employment effects of automation depend less on the technology itself and more

Preface

The global automotive industry is currently navigating one of the most profound periods of disruption in its history. A sector that for over a century relied on the internal combustion engine (ICE) and a linear model of private ownership is now being reshaped by a confluence of powerful forces: rapid technological innovation, urgent environmental imperatives, shifting consumer preferences, and evolving regulatory landscapes. This transformation is often encapsulated by the acronym CASE—Connected, Autonomous, Shared, and Electric mobility. These four pillars represent not merely incremental improvements but a fundamental paradigm shift that challenges the very core of how value is created, delivered, and captured in the mobility ecosystem.

We stand at a critical juncture where the decisions made today by original equipment manufacturers (OEMs), policymakers, technology providers, and consumers will define the trajectory of transportation for decades to come. The transition to a sustainable and intelligent mobility system is no longer a distant vision but an immediate strategic necessity. The urgency of climate change, underscored by international commitments such as the Paris Agreement and the UN Sustainable Development Goals (SDGs), has accelerated the push for decarbonization. Simultaneously, the digitalization of the economy has introduced new competitors and business models, forcing legacy automakers to pivot from being mere hardware manufacturers to becoming providers of integrated mobility services.

The shift is further complicated by geopolitical tensions and trade disputes, particularly between major economic powers like the European Union, China, and the United States, which are vying for leadership in the new energy vehicle (NEV) market. As supply chains are redrawn and protectionist measures rise, the industry must not only innovate technologically but also navigate a complex web of international relations and regulatory standards.

This book, *Connected, Autonomous, Shared, and Electric Vehicles in the New Age of Mobility*, emerges as a timely and essential resource to navigate this complex landscape. It offers a comprehensive examination of the multifaceted changes sweeping the industry, moving beyond a purely technological focus to explore the strategic, organizational, social, and human dimensions of this transition. By bringing together diverse perspectives—from dynamic capabilities and supply chain resilience to consumer psychology, workforce transformation, and socio-technical transitions—this volume provides a holistic roadmap for understanding the future of mobility.

OBJECTIVE OF THE BOOK

The primary objective of this publication is to provide a rigorous, evidence-based analysis of the strategic transformation currently underway in the automotive and transport sectors. It aims to bridge the gap between theoretical frameworks and practical application, offering actionable insights for stakeholders grappling with the challenges of CASE mobility.

Specifically, this book seeks to:

- **Analyze the Convergence of Forces:** To explore how electrification, connectivity, automation, and shared mobility are interacting to create new ecosystems, rather than operating in isolation.
- **Evaluate Strategic Responses:** To examine how organizations are leveraging dynamic capabilities, open innovation, and strategic partnerships to survive and thrive in a volatile environment.
- **Address the Sustainability Imperative:** To critically assess the role of circular economy principles, green supply chains, and sustainability certifications in reducing the environmental footprint of the sector.
- **Understand the Human and Social Element:** To delve into the changing behaviors of consumers, the profound impact of Industry 4.0 technologies on the automotive workforce, and the determinants of customer satisfaction in the EV era.
- **Contextualize Global Transitions:** To provide a macro-level view of the socio-technical transition from ICE to EVs, considering the role of government policy, trade dynamics, and global benchmarks like the Norwegian model.

By synthesizing cutting-edge research with real-world case studies, this book serves as a vital reference for those seeking to understand not just *what* is changing in the automotive world, but *why* and *how* organizations can successfully navigate this new age.

TARGET AUDIENCE

This book is designed to serve a broad yet specialized audience, reflecting the interdisciplinary nature of the mobility transition:

- **Academics and Researchers:** Scholars in the fields of strategic management, innovation, supply chain management, sociology of work, and sustainability will find valuable theoretical contributions and empirical data to support further research.
- **Industry Practitioners and Executives:** Managers and leaders within the automotive, technology, and energy sectors will benefit from the strategic frameworks and best practices outlined in the chapters, helping them make informed decisions in a high-stakes environment.
- **Policymakers and Regulators:** Government officials and urban planners tasked with designing the regulatory infrastructure for future mobility—from EV charging networks to autonomous vehicle safety standards—will find critical insights into the implications of these technologies.
- **Students:** Graduate and undergraduate students in business, engineering, and environmental studies will use this text to gain a deep understanding of the systemic challenges and opportunities facing one of the world's most vital industries.

ORGANIZATION OF THE BOOK

The book is organized into **8 chapters**, each addressing a critical dimension of the mobility transformation. The progression of chapters moves from a macro-level analysis of industry trends and strategic capabilities to specific technological enablers, business models, and finally, the human and social implications of these changes.

Chapter 1, "Strategic Transformation and Sustainability in CASE Mobility: Redefining Business Models and Value Chains for the Electric and Autonomous Future," sets the stage by defining the scope and scale of the CASE mobility revolution. It provides a foundational overview of the global mobility ecosystem, highlighting the exponential growth of the connected and electric vehicle markets. The chapter explores how the industry is shifting from a product-centric model to a service-oriented one, driven by the need for "zero ownership, zero emissions, and zero accidents." It introduces the dual lens of strategic transformation and value chain redesign, arguing that sustainability is no longer a compliance burden but a strategic imperative. This chapter frames the subsequent discussions by establishing that the vehicle of the future is not just a machine, but a "real-time connected digital experience platform."

Chapter 2, "Emerging Technologies Driving Sustainability in Automotive Supply Networks," focuses on the technological enablers of a transparent and ethical supply chain. It examines the lucrative potential of Artificial Intelligence (AI) and Blockchain in transforming Environmental, Social, and Governance (ESG) compliance. The chapter addresses critical issues such as data fragmentation and lack of traceability in global supply networks. Through case studies, it demonstrates how Blockchain can provide immutable records of sourcing and emissions, while AI forecasts sustainability threats. This chapter proposes a roadmap for incorporating these digital tools to create responsible, sustainable, and smart automotive networks, crucial for meeting the demands of regulators and ethically conscious consumers.

Chapter 3, "Sustainability Certifications: Tools for the Transformation of Industry," examines the role of third-party validations as instruments of industrial change. As stakeholders increasingly hold businesses accountable, certifications like ISO 14001, LEED, and Fair Trade have evolved from optional marketing badges to essential requirements for market legitimacy. This chapter details how these frameworks drive operational efficiency, foster innovation, and build consumer trust. It provides a typology of certifications—environmental, social, and product-focused—and illustrates through case studies how they compel organizations to institutionalize sustainability. The chapter argues that certifications are vital strategic tools for differentiation in a crowded market, signaling genuine commitment to global sustainability goals.

Chapter 4, "Decoding the Drive to Sustainability: A Customer-Centric Analysis of Sustainable Vehicle Adoption in a Key European Market," s The study delves on the antecedents of sustainable vehicle adoption (electric and hybrid cars) in Spain. Employing the Comprehensive Action Determination Model (CADM) as a theoretical framework, it accounts for four types of psychosocial factors, such as intentional variables, normative influences, situational variables, and habitual behaviour. Additionally, sociodemographic factors and pro-environmental behaviours are accounted for to obtain a more complete profile of the owner. After analysing the data from 358 individuals, the results demonstrate that the CADM model effectively explains the adoption of sustainable cars, indicating that intention and perceived behavioural control are the primary determinants of sustainable car adoption. Additionally, significant differences among (non)owners are observed across sociodemographic, psychographic, and behavioural factors. Implications and practical recommendations are provided, emphasising the necessity for policies and initiatives to address psychosocial barriers and facilitate transition towards environmentally sustainable mobility.

Chapter 5, "Sustainable AI Solution for the Transport System," addresses the systemic challenges of urbanization, congestion, and pollution through the lens of Artificial Intelligence. It provides a comprehensive review of AI applications in transportation, from predictive maintenance and traffic management to autonomous driving and logistics optimization. The chapter highlights how AI can reduce energy consumption and carbon emissions by optimizing routes and traffic flows in real-time. Furthermore, it explores frontier technologies like the Hyperloop, positioning AI not just as a tool for efficiency but as a foundational element for the next generation of sustainable transport infrastructure.

Chapter 6, "Automation and Digitalization in the Automotive Industry: Employment Impacts From the Case of Toyota in Mexico," This study analyzes the impact of automation and digitalization on employment through the lens of Skill-Biased Technological Change (SBTC), using the Toyota Guanajuato plant (TMMGT) as a case study. Using a qualitative methodology, the study identified an organizational strategy centered on internal relocation, continuous training, and upward mobility. This has allowed for the integration of new technologies without generating labor disruptions. The findings show that the employment effects of automation depend less on the technology itself and more on the organizational decisions that accompany its implementation. The TMMGT case exemplifies how a corporate culture focused on continuous improvement and human development can mitigate the exclusionary effects predicted by SBTC theory, enabling a more inclusive technological transition. This study contributes to the literature on digital transformation and employment in emerging economies. Future research should explore how organizational practices and institutional frameworks mediate the effects of technological change on work.

Chapter 7, "Policy Divergence in China's EV Market: The Cases of Hangzhou and Tianjin," This chapter investigates the significant divergence in EV market outcomes among Chinese cities, despite operating under a uniform national policy framework. Through a comparative case study of Hangzhou and Tianjin, the authors explain why two cities with similar socioeconomic profiles exhibit starkly different market penetration rates. The analysis employs an adapted policy mix framework, deconstructing each city's strategy into three key themes: demand-side regulation, supply-side industrial policy, and infrastructure governance. The findings reveal that Hangzhou's success is attributable to a coherent and consistent policy portfolio where stringent non-fiscal regulations create strong consumer demand. In contrast, Tianjin's slower adoption is linked to a less consistent policy mix constrained by path dependency on legacy automotive joint ventures. The chapter concludes that local state capacity to design and implement a synergistic policy mix is a critical determinant of success in sustainability transitions.

Chapter 8, "Evaluating Customer Satisfaction with Electric Vehicle (EV) Service and Quality," concludes the volume with a targeted empirical study that brings the discussion back to the end-user. As EVs move from niche to mainstream, understanding the drivers of customer satisfaction becomes paramount. This chapter investigates the impact of service quality, vehicle quality, pricing, and customer experience on overall satisfaction among EV owners. Through quantitative analysis, it reveals which factors are most significant in shaping consumer perceptions. This study provides critical feedback for manufacturers and service providers, emphasizing that widespread EV adoption relies not just on environmental altruism, but on superior product quality and service excellence that rivals or exceeds the ICE experience.

IMPACT AND IMPORTANCE

The transition to the new age of mobility is fraught with uncertainty, but it is also replete with opportunity. This book is significant because it refuses to view these changes in isolation. By weaving together the threads of technology, strategy, and society, it offers a blueprint for a transition that is not only technologically advanced but also socially equitable and environmentally sustainable.

For the academic community, this volume contributes to the theoretical understanding of dynamic capabilities and socio-technical transitions. For practitioners, it offers a pragmatic guide to the tools—from AI to certifications—that will define competitive advantage in the coming years. And for society at large, it articulates a vision of mobility that is cleaner, safer, and more accessible.

CONCLUSION

As we stand on the precipice of this new era, it is clear that the automotive industry is no longer just about building cars; it is about building the future of movement. The convergence of connectivity, autonomy, sharing, and electrification offers a path to resolve some of the most pressing challenges of our time. However, success is not guaranteed. It requires the deliberate alignment of technology with human needs and planetary boundaries.

We hope that *Connected, Autonomous, Shared, and Electric Vehicles in the New Age of Mobility* serves as both a compass and a catalyst for this journey, inspiring readers to engage deeply with the complexities of this transformation and to contribute actively to the creation of a sustainable mobility ecosystem.

Chapter 1
Strategic Transformation and Sustainability in CASE Mobility:
Redefining Business Models and Value Chains for the Electric and Autonomous Future

Bhavya Bhasuran
https://orcid.org/0009-0000-2299-1405
Sathyabama Institute of Science and Technology, India

Ashwin Prabhu G.
https://orcid.org/0000-0002-1614-7096
St. Joseph's College of Engineering, India

S. Karthik
Kalasalingam Academy of Research and Education, India

V. Naveen
St. Joseph's College of Engineering, India

R. Prasanth
St. Joseph's College of Engineering, India

ABSTRACT

This chapter explores the strategic transformation and sustainability imperatives shaping the evolution of Connected, Autonomous, Shared, and Electric (CASE) mobility. It examines how digitalization, decarbonization, and new business models

DOI: 10.4018/979-8-3373-4495-9.ch001

are redefining automotive value chains and enabling a shift from product-centric to service-oriented ecosystems. Integrating theoretical perspectives such as dynamic capabilities, the resource-based view, and circular economy principles, the chapter analyzes how organizations leverage innovation, data, and collaboration to achieve sustainable competitiveness. Through global case studies of leading mobility pioneers, it highlights emerging opportunities, policy enablers, and challenges in realizing sustainable, resilient, and inclusive mobility ecosystems. Ultimately, it emphasizes the need for integrated strategies that align technological advancement with environmental stewardship and socio-economic value creation in the new age of mobility.

1. INTRODUCTION

The global mobility ecosystem is undergoing a profound transformation. At the heart of this shift lies the convergence of four foundational trends—connectivity, autonomy, sharing, and electrification—often encapsulated by the acronym CASE mobility (Connected, Autonomous, Shared, and Electric). As noted by Puga Garcia (2023), the new mobility paradigm is no longer limited to individual car ownership, but is increasingly defined by digital platforms, shared services, electric propulsion and autonomous functionality. These developments are not merely technical upgrades; they signal a structural re-shaping of how value is created, delivered and captured across the entire automotive and mobility value chain. Industry sources estimate that the CASE mobility market is poised for exponential growth. For example, research by MarkWide Research forecasts that the global CASE market will reach USD 1.6 trillion by 2030, driven by rapid adoption of connected, autonomous, shared and electric vehicle technologies (MarkWide Research, 2023). Similarly, The Tech Outlook projects that by 2030 nearly 96% of new vehicles will possess built-in connectivity and that nearly one-quarter of mobility revenues will derive from new sources such as on-demand mobility (The Tech Outlook, 2023). These figures underscore the urgency for industry players to re-think their strategic positions and business models. At the same time, this transformation is deeply intertwined with the sustainability agenda. The mobility sector remains a major contributor to greenhouse gas emissions and urban air pollution, while new usage models (sharing rather than ownership), and electrification offer pathways to decarbonize and achieve 'zero-emissions' goals (Chirumalla, Klaff, Zako & Sannö, 2023). Moreover, the shift from product-centric to service-oriented models emphasises lifecycle thinking, circular economy principles and value-chain redesign (Tata Elxsi, 2024; Chakravarti, 2024). In sum, sustainability is no longer a peripheral compliance theme—it has become a strategic imperative for mobility ecosystem participants. This chapter adopts a dual lens: it explores strategic transformation within the automotive and mobility industry, and examines how sustainability imperatives are reshaping value chains in the age of CASE mobility. Strategic transformation is required because legacy automotive firms, Tier-1 and Tier-2 suppliers, infrastructure providers and new mobility entrants all face converging pressures: decreasing internal-combustion-engine (ICE) vehicle volumes, rising software-defined architectures, burgeoning data-driven services, and new regulatory mandates. As emphasised in the white paper from Tata Consultancy Services (TCS), the vehicle of the future will be "a real-time connected digital experience platform", rather than simply a machine for transport (TCS, 2024). Transforming towards that model demands new capabilities, ecosystems and business models. Value-chain redefinition is equally crucial. Traditional linear value chains—raw materials, components, assembly, distribution, sales, after-sales—are being disrupted. New mobility paradigms call for interconnected platforms, data-centric business models, circular supply chains (battery recycling, second-life modules), and mobility-as-a-service (MaaS) ecosystems. As outlined by Neckermann Strategic Advisors (2020), the mobility revolution is ushering in what they term "zero ownership, zero emissions, zero accidents," which necessitates a new value chain rather than simply extending the old one. The shift implies that the locus of value moves from the vehicle hardware to software, data, connectivity and service orchestration. From a sustainability perspective, the value-chain transformation is multi-dimensional. It involves environmental sustainability (electrification, decarbonisation, resource circularity), social sustainability (mobility access, job transitions, supply-chain transparency) and economic sustainability (new revenue streams, resilience in disrupted markets). For instance, recent studies emphasise that sharing and autonomous electric vehicles in B2B applications can generate novel business models driving circular economy logic (Chirumalla et al., 2023). Given this context, the primary aim of this chapter is to analyze how strategic transformation is being undertaken within the mobility sector in response to the rapid emergence of Connected, Autonomous, Shared, and Electric (CASE) trends, while also exploring how sustainability imperatives are being integrated into value chain redesign across the evolving mobility ecosystem. It further seeks to provide comprehensive insights and conceptual frameworks that help practitioners and researchers align strategic transformation efforts with sustainable value chain models to ensure long-term competitiveness and environmental responsibility. By merging strategic management perspectives such as the dynamic capabilities and resource-based view with sustainability paradigms including the circular economy and triple bottom line, the discussion bridges theoretical understanding with real-world applications. Ultimately, the era of CASE mobility represents not merely an incremental technological shift but a fundamental reimagination of the global mobility ecosystem. Organizations that thrive in this transition will be those that reconceptualize the vehicle not as a standalone product but as part of a connected, shared, electrified, and intelligent service network—one that demands strategic transformation, reconfigured value chains, and a deep commitment to sustainability at its core.

2. BACKGROUND AND EVOLUTION OF CASE MOBILITY

The global automotive and transport sector is experiencing a structural transformation shaped by the convergence of four forces—Connected, Autonomous, Shared, and Electric mobility (CASE). This shift is redefining industrial value chains, consumer expectations, and sustainability priorities. Traditional mobility, centered on internal-combustion vehicles and ownership, is giving way to digitalized, service-based, and eco-conscious ecosystems that integrate smart technologies with low-carbon objectives (Huang, Kockelman, & Gurumurthy, 2022). Connectivity serves as the digital foundation of CASE. Through vehicle-to-everything (V2X) communication, telematics, and cloud analytics, vehicles now function as intelligent nodes that exchange data with infrastructure and users. These systems enable predictive maintenance, software updates, and adaptive routing, while contributing to accident prevention and traffic optimization (Huang et al., 2022). The ability to collect and monetize operational data is creating new revenue channels and altering competition from manufacturing efficiency to digital service innovation. Automation represents the next frontier. Autonomous vehicles (AVs) use sensors, control algorithms, and machine learning to navigate without human intervention. Guanetti, Kim, and Borrelli (2018) describe how these systems promise safer and more inclusive transport, though they raise technical, ethical, and legal challenges. As autonomy levels advance, the vehicle becomes a software-defined platform. When combined with connectivity, fleets of connected autonomous vehicles (CAVs) can coordinate in real time, optimizing routing and reducing congestion. This evolution signals a fundamental reorientation of automotive business models toward software and artificial-intelligence capabilities rather than mechanical differentiation. The shared-mobility dimension challenges the long-standing ownership paradigm. Ride-sharing, car-sharing, and subscription schemes have proliferated, driven by urbanization, digital platforms, and changing social preferences. Mobility-as-a-Service (MaaS) systems integrate various transport options—public transit, shared vehicles, and micromobility—into unified apps that manage booking and payment. These models increase vehicle utilization, decrease congestion, and reduce emissions (Netsol Tech, 2024). Data integration and user experience design thus become central sources of competitive advantage.

Electrification forms the environmental pillar of CASE. Electric vehicles (EVs) are crucial for decarbonizing transport, which currently contributes nearly one-quarter of global greenhouse-gas emissions (International Energy Agency [IEA], 2023). The shift to batteries or fuel cells requires new supply chains, charging infrastructure, and recycling systems. Governments worldwide promote this transition through incentives and carbon-neutrality policies. Nevertheless, sustainability concerns persist regarding battery materials and disposal, prompting circular-economy approaches such as reuse and recycling (Netsol Tech, 2024). Electrification therefore demands not only technological innovation but also responsible resource management. The integration of connectivity, autonomy, sharing, and electrification generates synergies that exceed the sum of their parts. Connectivity supplies real-time data; autonomy interprets it; sharing optimizes vehicle use; and electrification ensures climate alignment. Collectively, these forces produce a networked mobility ecosystem emphasizing data, software, and sustainability. The Tech Outlook (2023) projects that by 2030, most new vehicles will include embedded connectivity, and a significant share of industry revenue will derive from digital and on-demand services rather than physical sales. MarkWide Research (2023) similarly anticipates that the global CASE market will reach USD 1.6 trillion by 2030, illustrating how value creation is shifting toward data management, software platforms, and integrated services. This transformation is also altering the industry's stakeholder structure. Original equipment manufacturers (OEMs) now collaborate with technology companies, utilities, and infrastructure providers. The Center for Automotive Research (C.A.R. Group, 2021) notes that the linear supply chain is evolving into a complex ecosystem where partnerships and innovation networks determine success. Cross-sector alliances among automakers, telecom operators, and software firms are crucial to deliver connected and autonomous functionalities. Such collaboration blurs traditional industry boundaries and introduces new governance, cybersecurity, and intellectual-property challenges.

Sustainability remains the central motivation behind CASE mobility. Global regulations—such as the EU Green Deal, India's FAME II program, and the U.S. Inflation Reduction Act (2022)—encourage cleaner technologies and infrastructure investment. Yet sustainability extends beyond electrification to include ethical sourcing, life-cycle design, and workforce adaptation. Tan and Leon-Garcia (2018) emphasize that future smart cities will rely on autonomous and energy-managed systems integrated with renewable grids. Achieving sustainable mobility thus requires embedding circular-economy practices, protecting data privacy, ensuring equitable access, and preparing workers for digital transitions. Ultimately, CASE mobility represents a paradigm shift from mechanical engineering toward a digital, service-oriented, and environmentally responsible model. It transcends the conventional concept of the automobile to establish a connected ecosystem where vehicles, infrastructure, and users operate symbiotically. For industry stakeholders, success depends on mastering data analytics, software innovation, and sustainability integration. The convergence of these technologies promises safer, cleaner, and more inclusive transport, marking the dawn of a redefined and intelligent mobility era that aligns economic growth with planetary stewardship.

3. STRATEGIC TRANSFORMATION IN THE MOBILITY ECOSYSTEM

The mobility industry is undergoing a paradigm shift driven by the convergence of digitalization, sustainability imperatives, and evolving consumer expectations under the Connected, Autonomous, Shared, and Electric (CASE) mobility framework. Strategic transformation in this sector extends beyond technological innovation—it redefines organizational structures, leadership, and value creation models. Firms are increasingly required to develop dynamic capabilities, embrace digital ecosystems, and engage in strategic partnerships to remain competitive in this rapidly changing landscape.

3.1 Digital Transformation as a Strategic Imperative

Digital transformation represents the cornerstone of strategic change in the automotive and transportation industries. Historically, automakers operated in a product-centric paradigm, emphasizing efficiency in manufacturing and cost optimization. However, CASE mobility requires organizations to evolve into digital platforms that integrate data analytics, artificial intelligence (AI), and cloud computing to deliver mobility services rather than just physical products (Bohnsack et al., 2020). AI-powered analytics and intelligent systems play a pivotal role in predictive maintenance, demand forecasting, and resource optimization in connected mobility ecosystems (Dhanushya et al., 2025). Data has emerged as a critical strategic resource. Vehicle-to-everything (V2X) connectivity enables the collection of massive datasets from vehicles, infrastructure, and users, facilitating predictive maintenance, personalized services, and real-time traffic management (KPMG, 2023). Firms such as Tesla and BYD exemplify this transformation by integrating software-driven architectures and over-the-air (OTA) updates, enabling continuous product improvement and customer engagement (Deloitte, 2022). Digital platforms are not only redefining business operations but also transforming value perception. Consumers increasingly value access over ownership, preferring mobility services that provide convenience, sustainability, and seamless integration with digital lifestyles (Roland Berger, 2022). Consequently, traditional automakers are investing in mobility-as-a-service (MaaS) platforms and subscription models that emphasize usage-based value generation. This transformation requires deep integration of information technology and operational systems, challenging firms to balance legacy infrastructures with emerging digital ecosystems.

3.2 Developing Dynamic Capabilities in a Rapidly Evolving Environment

Dynamic capabilities—the ability of firms to integrate, build, and reconfigure internal and external competences to address rapidly changing environments—are central to strategic transformation in CASE mobility (Teece, 2018). The speed and uncertainty associated with technological disruption demand organizational agility, innovation ambidexterity, and proactive learning mechanisms. Advances in FPGA-based adaptive systems demonstrate how real-time data correction and noise elimination can enhance vehicular sensor accuracy and diagnostic efficiency (Shanmugaraja et al., 2023). Automotive leaders such as BMW, Hyundai, and Toyota have reorganized around agile teams and modular architectures to enhance flexibility and responsiveness. BMW's "Strategy NEXT" initiative integrates digital intelligence, electrification, and sustainability goals through modular platforms that support cross-functional collaboration (BMW Group, 2023). Similarly, Toyota's investment in autonomous driving technology through its subsidiary Woven Planet underscores its commitment to continuous learning and innovation ecosystems. Autonomous control and real-time decision-making, as demonstrated in intelligent rover and robotics systems, parallel the principles driving connected and self-driving vehicles (Ruba et al., 2023). AI-assisted inspection and fault detection mechanisms can significantly enhance vehicle and infrastructure reliability in intelligent transport systems (Ruba et al., 2025).

Dynamic capabilities also manifest in adaptive leadership and talent management. Firms must cultivate a workforce capable of interdisciplinary collaboration between software engineering, data analytics, and mechanical design. The shift toward software-defined vehicles has expanded the demand for digital skills and redefined automotive engineering roles (Accenture, 2022). Leadership now requires digital fluency and a systems-thinking approach that aligns corporate strategy with technological advancement and environmental responsibility.

3.3 Strategic Partnerships and Ecosystem Collaboration

In the CASE era, no single organization can independently master all technologies or infrastructure requirements. Collaboration across industries—automotive, telecommunications, energy, and software—has become indispensable. Strategic alliances enable firms to share risks, accelerate innovation, and scale operations (McKinsey & Company, 2022). For instance, alliances such as Ford and Google's data partnership leverage AI and cloud capabilities to enhance customer experiences and operational efficiency. Similarly, Volkswagen's collaboration with Argo AI illustrates how automakers are co-developing autonomous vehicle technologies with specialized tech startups (Volkswagen Group, 2022). These partnerships reflect a transition from vertical integration to networked ecosystems, where value is co-created among diverse stakeholders. As mobility systems become more interconnected, identifying and mitigating cyber threats through machine learning and pattern recognition becomes vital (Krithika et al., 2025).

Governments and urban authorities also play crucial roles as ecosystem partners by investing in smart infrastructure, charging networks, and regulatory frameworks. Public-private partnerships (PPPs) have been instrumental in accelerating electric vehicle (EV) adoption in regions like Europe and Asia (International Energy Agency [IEA], 2023). The integration of energy utilities, transport planners, and digital service providers creates systemic synergies essential for sustainable mobility ecosystems. However, ecosystem collaboration also presents challenges. Issues related to data ownership, cybersecurity, and governance must be carefully managed. The success of partnerships depends on clear contractual frameworks, trust-based governance, and interoperable digital standards (PwC, 2023). As ecosystems expand, firms must adopt open innovation models that balance competitiveness with collective value creation.

3.4 Business Model Innovation and Value Chain Reconfiguration

The transformation of mobility from a product-based to a service-based model requires radical business model innovation. Traditional automakers focused on manufacturing and aftersales services, but the CASE paradigm emphasizes lifecycle management, data monetization, and user-centric experiences (Bohnsack et al., 2020). Business models such as mobility-as-a-service (MaaS), battery-as-a-service (BaaS), and vehicle subscription programs redefine how value is captured. For instance, NIO's BaaS model allows consumers to lease batteries separately from vehicles, reducing upfront costs and enhancing flexibility (NIO, 2022). This demonstrates how decoupling hardware and software components can create scalable revenue streams.

The transition also impacts value chains. Digitalization and sustainability imperatives are reshaping supplier relationships and creating new intermediaries such as software integrators, data service providers, and charging network operators (Roland Berger, 2022). Traditional tier-based supplier hierarchies are giving way to platform-based ecosystems where data flow and digital interfaces determine value capture. Moreover, electrification and circular economy practices are pushing firms to redesign supply chains to ensure sustainable sourcing, recycling, and reuse of materials (World Economic Forum, 2023). The integration of blockchain technology is being explored to enhance transparency and traceability in EV battery supply chains (PwC, 2023). Thus, strategic transformation involves not only technological adoption but also structural realignment toward resilient, sustainable, and digitally connected value networks.

3.5 Leadership for Transformative Change

Successful strategic transformation in the mobility ecosystem requires visionary leadership capable of orchestrating cultural change and managing cross-sector complexity. Leaders must embrace long-term sustainability goals, foster innovation, and empower decentralized decision-making. The ability to balance exploration (innovation and experimentation) with exploitation (efficiency and scale) determines organizational resilience (Teece, 2018). Transformative leadership involves cultivating a culture of agility and inclusivity. As organizations diversify their talent base to include digital and sustainability experts, leaders must promote continuous learning and psychological safety to encourage experimentation (Accenture, 2022). Governance structures should align sustainability performance with executive incentives, ensuring that environmental and social metrics are embedded in strategic evaluation frameworks. Ultimately, strategic transformation in CASE mobility is not a one-time initiative but a continuous evolutionary process. It demands adaptive leadership, ecosystem collaboration, and business model flexibility to navigate technological uncertainty and societal expectations. Organizations that integrate digital innovation, sustainability, and strategic foresight will be best positioned to thrive in the future mobility landscape.

4. REDEFINING VALUE CHAINS FOR CASE MOBILITY

The emergence of Connected, Autonomous, Shared, and Electric (CASE) mobility has disrupted traditional automotive value chains and created an interconnected web of digital, sustainable, and service-oriented ecosystems. Unlike the conventional linear supply chains centered on manufacturing efficiency and component integration, the new mobility value chain emphasizes software, data analytics, and cross-sector collaboration. The transformation reflects a systemic shift from mechanical production to digital orchestration, from ownership to service, and from fossil-based energy to renewable integration. This reconfiguration is redefining how value is created, captured, and distributed across the global mobility landscape. Organizational agility and interdisciplinary digital strategies are crucial in adapting to the fast-paced transformation of CASE ecosystems (Ruba et al., 2026a).

4.1 From Linear to Ecosystem-Oriented Value Chains

Traditional automotive value chains followed a linear, tiered structure: original equipment manufacturers (OEMs) designed and assembled vehicles, while suppliers provided standardized components in hierarchical arrangements. This model relied heavily on economies of scale, long-term contracts, and vertically integrated production (Sturgeon & Van Biesebroeck, 2011). However, the CASE paradigm has dismantled these rigid hierarchies. The convergence of digital platforms, renewable energy systems, and shared mobility services requires collaborative ecosystems involving automakers, software firms, telecom providers, infrastructure developers, and government bodies (McKinsey & Company, 2022). Ecosystem-based value chains operate on interdependence rather than control. Value creation occurs through platform-mediated interactions and co-development of services. For instance, Tesla's vertical integration combines vehicle manufacturing, software updates, and energy services, but its ecosystem also includes partnerships with charging networks and battery suppliers (Deloitte, 2022). Similarly, companies like Uber and Waymo operate as mobility platform orchestrators, leveraging data from connected vehicles and users to create digital marketplaces. These platforms serve as intermediaries linking supply (drivers or fleets) and demand (users or logistics partners), illustrating the transition from product-centric to network-centric value generation.

4.2 The Rise of Software-Defined Mobility

Software now represents the most valuable layer of the automotive value chain. In a CASE environment, vehicles are effectively "computers on wheels," where hardware serves as a platform for software-driven innovation. Over-the-air (OTA) updates, advanced driver-assistance systems (ADAS), and cloud-based analytics enable continuous improvement of vehicles post-sale (Roland Berger, 2022). The value once derived from physical production is increasingly replaced by digital services, subscription models, and data monetization. Automakers such as General Motors and Mercedes-Benz are reengineering their value chains around software architectures that support connectivity, automation, and customer personalization (Accenture, 2023). This software-defined transformation also encourages modular production systems that separate hardware from software, allowing faster integration of new features without redesigning the vehicle structure (KPMG, 2023). The monetization of mobility data—from route optimization to driver behavior insights—introduces new competitive advantages and potential ethical challenges related to privacy and data governance (PwC, 2023).

Software-defined mobility also expands the boundaries of traditional automotive value creation. Partnerships with AI firms, cloud providers, and cybersecurity specialists have become vital components of modern supply chains. The integration of AI-driven predictive maintenance, intelligent navigation, and connected infotainment exemplifies how digital ecosystems are embedding intelligence into every layer of the mobility experience (McKinsey & Company, 2022). The resulting structure resembles a digital "value web," where feedback loops and data flows continuously reshape market relationships.

4.3 Sustainability as a Structural Driver of Value Chain Redesign

Sustainability has become an inseparable driver of value chain reconfiguration in the mobility industry. Governments, investors, and consumers increasingly demand that automotive and transport systems adhere to environmental, social, and governance (ESG) principles (World Economic Forum, 2023). Electrification, renewable energy integration, and circular economy strategies are redefining material sourcing, production, and product lifecycle management. Electric vehicle (EV) production illustrates this transformation vividly. Traditional internal combustion engine (ICE) supply chains relied on thousands of mechanical components, while EVs use far fewer parts but require new materials such as lithium, cobalt, and nickel for batteries (International Energy Agency [IEA], 2023). The sourcing of these critical minerals introduces sustainability and ethical concerns, compelling firms to adopt transparent and traceable procurement systems. Blockchain technology is increasingly deployed to ensure responsible sourcing and manage battery lifecycle data (Deloitte, 2023).

Circular economy principles are also reshaping manufacturing and recycling strategies. Firms like Renault and Volvo are designing vehicles with recyclability in mind, implementing remanufacturing facilities that extend product life and reduce waste (World Economic Forum, 2023). Battery recycling partnerships—such as Toyota's collaboration with Redwood Materials—illustrate the growing importance of reverse logistics and closed-loop supply chains. These practices not only reduce environmental impact but also enhance economic resilience by recovering valuable materials from end-of-life products. Sustainability in value chains also encompasses the shift toward renewable-powered logistics and production. OEMs and suppliers are increasingly decarbonizing operations through renewable energy sourcing, energy-efficient factories, and green logistics networks (Accenture, 2023). This integration aligns mobility with the broader goals of the UN Sustainable Development Goals (SDGs), particularly those focused on climate action (SDG 13) and responsible production (SDG 12).

4.4 Strategic Partnerships and Platform Collaboration

The complexity of CASE value chains necessitates inter-organizational collaboration. Unlike traditional supplier-buyer relationships, these partnerships are multi-directional and dynamic. Strategic alliances between automakers and technology providers have become the norm, allowing firms to pool expertise and share innovation costs. For instance, Ford's collaboration with Google to leverage AI-driven data analytics, and Stellantis's partnership with Amazon Web Services (AWS) for connected car development, highlight the emergence of "mobility data ecosystems" (McKinsey & Company, 2022). Energy companies, telecom operators, and digital service firms are also integral to these redefined value chains. The deployment of EV charging infrastructure relies heavily on partnerships among utilities, charging-network providers, and governments (IEA, 2023). Telecom firms such as Ericsson and Huawei are enabling the communication backbone for autonomous and connected vehicles, integrating 5G connectivity to support real-time operations. These cross-sectoral collaborations illustrate how value is increasingly co-created among interdependent players.

The challenge lies in balancing openness and competitiveness. Ecosystem governance models must ensure interoperability, cybersecurity, and equitable data sharing. Standardization of software interfaces and data protocols is essential for maintaining seamless interoperability across platforms. Regulatory bodies in the EU, North America, and Asia are promoting common standards for EV charging, vehicle-to-grid (V2G) integration, and autonomous vehicle testing, thereby shaping global value chain architectures (KPMG, 2023).

4.5 Value Capture and the Emerging Mobility Economy

The redefinition of mobility value chains has also altered mechanisms of value capture. In traditional automotive models, value capture was concentrated at the point of sale, with limited post-purchase engagement. In contrast, CASE mobility emphasizes lifecycle revenue through digital services, data subscriptions, and platform participation. McKinsey (2022) estimates that digital and software-based services could account for up to 30% of total automotive industry revenues by 2030.

Service-based models such as mobility-as-a-service (MaaS), fleet management systems, and battery-leasing programs diversify income sources while enhancing customer loyalty. Automakers are transforming into service orchestrators, offering integrated solutions across transportation, energy management, and infotainment. This shift also redistributes value toward technology firms, software developers, and infrastructure providers—challenging traditional OEM dominance (PwC, 2023). Moreover, consumer trust and ethical responsibility are becoming critical elements of value capture. Companies that demonstrate transparent data practices, sustainable sourcing, and inclusive mobility design are gaining reputational and financial advantages. Thus, value in the new mobility ecosystem extends beyond profit to include social and environmental impact, aligning with the triple bottom line framework (Elkington, 1998).

4.6 Toward a Resilient and Regenerative Mobility Ecosystem

The ultimate goal of redefining value chains under CASE mobility is to build resilience and regenerative capacity. The COVID-19 pandemic, semiconductor shortages, and geopolitical tensions have underscored the vulnerability of global supply networks. In response, firms are regionalizing production, diversifying suppliers, and investing in digital twin technologies to simulate and optimize supply chain resilience (Accenture, 2023). A regenerative mobility ecosystem goes beyond resilience by actively restoring environmental and social systems. Through sustainable design, circularity, and equitable access, organizations can create self-sustaining value cycles that align profitability with planetary health. As mobility evolves into an integrated network of energy, data, and transport systems, redefining value chains becomes a strategic necessity for shaping a sustainable and inclusive future.

5. SUSTAINABILITY AND GREEN TRANSFORMATION IN CASE MOBILITY

Sustainability has become the cornerstone of modern mobility transformation, influencing technological innovation, industrial policy, and corporate strategy across the Connected, Autonomous, Shared, and Electric (CASE) paradigm. The mobility sector—long associated with carbon emissions, resource depletion, and environmental degradation—is now at the center of a systemic transition toward carbon neutrality and circularity. This shift is not only driven by regulatory pressure but also by consumer expectations, investor priorities, and the ethical responsibility of organizations to contribute to global climate goals. The sustainable transformation of mobility thus represents a convergence of technological innovation, environmental stewardship, and socio-economic equity.

5.1 The Strategic Imperative for Sustainable Mobility

The United Nations Sustainable Development Goals (SDGs), particularly SDG 7 (Affordable and Clean Energy), SDG 11 (Sustainable Cities and Communities), and SDG 13 (Climate Action), provide the global framework guiding mobility transformation (United Nations, 2022). The transportation sector accounts for approximately one-quarter of global energy-related CO_2 emissions, making decarbonization a critical priority (International Energy Agency [IEA], 2023). In response, governments and industries worldwide are committing to net-zero targets and mandating cleaner mobility technologies, such as electric vehicles (EVs) and renewable-powered public transportation.

Automotive firms are repositioning their strategies to align with sustainability goals. Companies like BMW, Toyota, and Volvo have established science-based targets to reduce life-cycle emissions, while newer entrants like Tesla and Rivian are integrating sustainability into their brand identities. Sustainability has evolved from a compliance-driven concern to a core strategic differentiator, shaping investments, partnerships, and consumer engagement (KPMG, 2023). In this context, sustainability acts not merely as an environmental objective but as a value creation mechanism, influencing product design, manufacturing, logistics, and end-of-life management.

5.2 Electrification and Renewable Integration

Electrification is the most visible pillar of the sustainable mobility transition. Electric vehicles reduce dependence on fossil fuels, promote energy diversification, and enable integration with renewable energy grids. Global EV sales surpassed 14 million units in 2023, representing nearly 18% of total car sales (IEA, 2023). Governments have accelerated adoption through incentives, carbon pricing, and infrastructure development, while private players are investing heavily in battery innovation and charging networks. Developing sustainability-oriented human capital through AI-integrated education is essential to ensure equitable and responsible technological adoption in CASE mobility (Santhini et al., 2025). Conversational AI platforms also contribute to stakeholder engagement and public awareness of sustainable practices in the mobility sector (Ruba et al., 2026b). However, electrification alone does not guarantee sustainability. The sourcing of critical minerals—such as lithium, cobalt, and nickel—raises environmental and ethical challenges related to mining practices and labor conditions (World Economic Forum, 2023). To address these issues, companies are investing in responsible sourcing frameworks and traceability systems using blockchain technology (Deloitte, 2023). Moreover, recycling and second-life applications for EV batteries are becoming vital components of the circular mobility ecosystem. Closed-loop battery value chains reduce dependency on virgin resources and minimize waste.

The integration of electric mobility with renewable energy systems further enhances sustainability outcomes. Vehicle-to-grid (V2G) technologies allow EVs to act as distributed energy storage units, stabilizing electricity grids and facilitating the use of solar and wind power (Accenture, 2023). This convergence of transport and energy sectors illustrates the emergence of "green mobility ecosystems," where clean power generation, digital energy management, and intelligent transportation coalesce.

5.3 Circular Economy and Resource Efficiency

The circular economy is reshaping the automotive industry's production and consumption models. Traditional linear systems—"take, make, dispose"—are being replaced by regenerative models emphasizing reuse, remanufacturing, and recycling. Circular mobility strategies aim to maximize material value, extend product lifespans, and minimize waste. For example, Renault's "Refactory" in France reconditions vehicles and components, reducing carbon emissions by up to 60% compared to producing new parts (World Economic Forum, 2023). Circular design also involves adopting modular vehicle architectures that facilitate component replacement and recycling. The use of biodegradable composites, recycled metals, and 3D-printed parts reduces environmental footprints while enhancing production flexibility (Roland Berger, 2022). Additionally, shared mobility platforms—such as car-sharing and ride-pooling services—contribute to circular principles by increasing vehicle utilization rates and reducing the total number of vehicles required to meet urban transport demand. This evolution from ownership to access signifies a behavioral and systemic transformation in mobility consumption.

End-of-life management remains a critical component of the circular ecosystem. Effective recycling of batteries, tires, and plastics requires coordination among manufacturers, recyclers, and policymakers. Emerging battery recycling firms such as Redwood Materials and Northvolt exemplify how circularity can create new industries and employment opportunities. Ultimately, circularity enhances both environmental and economic resilience, ensuring long-term resource security in an era of growing material scarcity.

5.4 Green Supply Chains and Carbon-Neutral Operations

Achieving sustainability in mobility extends beyond products to the entire supply chain. Green supply chains integrate low-carbon manufacturing, renewable energy use, and sustainable logistics practices. Many OEMs are now engaging suppliers through carbon accounting, renewable procurement, and sustainability reporting frameworks (Deloitte, 2022). Digital tools such as blockchain, IoT sensors, and AI-based analytics enable real-time monitoring of emissions and material flows, enhancing transparency and accountability. Carbon-neutral manufacturing has become a central goal. For example, Mercedes-Benz aims for carbon-neutral production across all plants by 2039, while BMW and Toyota have introduced solar and wind-powered factories (KPMG, 2023). Sustainable logistics solutions—like electric freight vehicles and route optimization algorithms—further reduce emissions across distribution networks. Moreover, partnerships with renewable energy providers facilitate the greening of supply operations, reinforcing the interconnectedness between the energy and transport sectors. Green transformation also requires rethinking workforce skills and stakeholder engagement. Sustainable mobility demands expertise in energy management, data analytics, and systems thinking. Organizations must therefore cultivate "green capabilities" through training and collaboration. Sustainability-oriented leadership and culture are emerging as critical enablers of long-term transformation (Accenture, 2023).

5.5 Toward an Equitable and Regenerative Mobility Future

The sustainability agenda in CASE mobility is incomplete without addressing equity and inclusion. Green transformation must ensure that technological progress benefits all segments of society, not just affluent consumers in urban centers. Accessibility to clean, affordable, and efficient mobility remains a global challenge, particularly in developing economies. Policies promoting inclusive mobility—such as subsidized EV programs, rural electrification, and accessible public transport—are vital for achieving social sustainability (United Nations, 2022). The future of sustainable mobility lies in regenerative systems—those that not only minimize harm but also restore ecological and social balance. Regenerative mobility envisions transport systems that contribute positively to the planet through renewable integration, zero-waste manufacturing, and equitable access. As mobility ecosystems evolve, the fusion of technological innovation, circular design, and ethical governance will define the path toward a sustainable and inclusive future.

6. THEORETICAL AND CONCEPTUAL MODELS FOR CASE TRANSFORMATION

The transition toward Connected, Autonomous, Shared, and Electric (CASE) mobility represents one of the most profound industrial transformations of the 21st century, integrating technological innovation, sustainability, and systemic change. To understand and navigate this complex evolution, a range of theoretical and conceptual models provide valuable analytical frameworks. These models—rooted in strategic management, innovation theory, and sustainability studies—explain how organizations build resilience, foster innovation, and realign value creation in response to disruptive mobility paradigms.

6.1 Dynamic Capabilities and Strategic Agility

Dynamic capabilities theory, introduced by Teece, Pisano, and Shuen (1997), provides a foundational perspective on how firms adapt to fast-changing environments. It emphasizes three key capacities: sensing opportunities, seizing them through innovation, and reconfiguring organizational assets to sustain competitiveness. In the context of CASE mobility, dynamic capabilities enable organizations to navigate technological turbulence, regulatory changes, and evolving consumer preferences. Energy-efficient hardware and memory optimization, such as precharge-free architectures, form a foundational aspect of low-latency computing in autonomous mobility applications (Kameswaran et al., 2017). For instance, automakers like General Motors and Hyundai have demonstrated strong sensing capabilities by anticipating the rise of electric and autonomous vehicles. Their investments in battery technologies and AI-based systems illustrate their ability to seize emerging opportunities (McKinsey & Company, 2022). Reconfiguration, on the other hand, is evident in the restructuring of supply chains and partnerships—where traditional OEMs collaborate with software firms and energy providers to co-create value within new ecosystems (KPMG, 2023). Thus, dynamic capabilities provide a robust lens to explain how strategic agility underpins transformation in the mobility industry.

6.2 Resource-Based View (RBV) and Digital Assets

The Resource-Based View (RBV) complements dynamic capabilities by focusing on internal competencies and resources that confer sustained competitive advantage. According to Barney (1991), valuable, rare, inimitable, and non-substitutable (VRIN) resources drive organizational success. In the CASE paradigm, these resources are increasingly digital, data-driven, and ecosystem-oriented. Traditional manufacturing excellence is no longer sufficient; firms must now leverage software capabilities, data analytics, and cloud infrastructure as core strategic assets. Tesla's vertically integrated data ecosystem and over-the-air software update capability exemplify how data and digital intelligence can become inimitable resources (Deloitte, 2023). Similarly, Volkswagen's "Cariad" software subsidiary represents a strategic attempt to internalize digital competencies and build proprietary data-driven platforms (Volkswagen Group, 2023). The RBV thus evolves in the CASE context to encompass digital capital, intellectual property, and partnerships as sources of enduring value.

6.3 Innovation Ecosystem and Open Innovation Models

Innovation in CASE mobility is inherently ecosystemic—requiring collaboration among automakers, technology firms, energy providers, governments, and consumers. The innovation ecosystem model (Adner, 2017) provides a theoretical framework to understand how interdependent actors jointly create and capture value. Unlike traditional linear innovation processes, ecosystem innovation involves co-specialization, complementarity, and network orchestration. For example, the development of autonomous driving systems demands collaboration across software developers, sensor manufacturers, AI researchers, and regulators. Companies like Waymo and Mobileye rely on open innovation ecosystems, leveraging partnerships to scale complex technologies (Roland Berger, 2022). The success of these systems depends not only on technological breakthroughs but also on governance mechanisms that coordinate multiple stakeholders' interests. Open innovation, as described by Chesbrough (2003), further strengthens this approach by encouraging firms to combine internal R&D with external knowledge flows. The mobility industry's partnerships with startups, universities, and public institutions illustrate how open innovation accelerates CASE transformation. By sharing risks and leveraging external expertise, firms can shorten development cycles and enhance adaptability in rapidly evolving technological landscapes.

6.4 Triple Bottom Line and Circular Economy Frameworks

Sustainability within CASE transformation cannot be separated from theoretical frameworks that integrate environmental, social, and economic dimensions. The Triple Bottom Line (TBL) model, proposed by Elkington (1997), provides a holistic lens for assessing business performance beyond profit—emphasizing "People, Planet, and Profit." In the mobility context, this means evaluating how vehicle production, energy consumption, and mobility access contribute to broader societal and ecological outcomes. For instance, mobility service providers like Uber and Bolt are incorporating TBL principles by committing to zero-emission operations and fair labor standards (World Economic Forum, 2023). Similarly, circular economy frameworks operationalize sustainability through resource efficiency, closed-loop systems, and life-cycle management. Concepts such as battery recycling, remanufacturing, and modular vehicle design directly align with circularity principles (Deloitte, 2022). Together, TBL and circular economy models redefine value creation as regenerative rather than extractive, positioning sustainability as a core driver of innovation.

6.5 Socio-Technical Systems Theory and Systemic Transformation

CASE mobility transformation is not merely technological—it is socio-technical. Socio-technical systems theory, derived from the work of Trist and Emery (1951), explains how technological change interacts with social, organizational, and institutional dimensions. The shift toward autonomous and shared mobility involves reconfiguring not only products and services but also user behaviors, regulatory frameworks, and cultural norms. For instance, the acceptance of self-driving vehicles depends on public trust, ethical governance, and social readiness, not just technical reliability (OECD, 2022). Similarly, shared mobility models require societal adaptation in ownership attitudes and privacy norms. By applying socio-technical systems thinking, policymakers and industry leaders can design mobility transitions that are both technologically feasible and socially sustainable. This integrated approach is vital for ensuring that innovation leads to equitable and human-centered outcomes.

6.6 Integrative Framework for CASE Transformation

Synthesizing the above theories reveals an integrative conceptual framework for understanding CASE transformation. Dynamic capabilities explain how firms adapt; RBV identifies the internal assets enabling this adaptation; innovation ecosystems and open innovation describe how collaboration accelerates value creation; TBL and circular economy models ensure environmental and social alignment; and socio-technical systems theory contextualizes change within broader human and institutional dynamics. Firms that strategically combine these frameworks can achieve not only competitive advantage but also long-term resilience and purpose-driven growth in the evolving mobility landscape.

7. CASE STUDIES AND INDUSTRY APPLICATIONS IN CASE MOBILITY

`The implementation of Connected, Autonomous, Shared, and Electric (CASE) mobility principles is reshaping global transportation systems through real-world applications that blend digital transformation, sustainability, and innovation. Industry leaders are reconfiguring value chains, experimenting with new business models, and deploying advanced technologies to achieve strategic agility and environmental goals. This section examines select case studies across major mobility domains—electric mobility, autonomous systems, shared platforms, and digital ecosystems—to illustrate how organizations are operationalizing CASE transformation and redefining the future of transportation.

7.1 Tesla: Digital Integration and Vertical Control in Electric Mobility

Tesla remains one of the most transformative forces in the mobility industry, pioneering the convergence of electrification, connectivity, and digital business models. Its vertically integrated strategy—encompassing battery production, software development, and energy services—demonstrates the power of combining product innovation with ecosystem control (Deloitte, 2023). Tesla's vehicles are essentially software platforms on wheels, equipped with over-the-air (OTA) update capabilities that continuously enhance performance and safety. This capability not only strengthens customer engagement but also creates a data-driven feedback loop that accelerates product improvement. From a value-chain perspective, Tesla's Gigafactories embody both circular economy and sustainability principles. By co-locating cell manufacturing, recycling, and assembly, Tesla minimizes logistics emissions and operational waste (World Economic Forum, 2023). Moreover, its integration with renewable energy—through Solar Roofs and Powerwalls—illustrates how electric mobility aligns with clean-energy ecosystems. The company's business model exemplifies dynamic capabilities: sensing opportunities in battery cost reduction, seizing them through innovation, and reconfiguring production networks to scale sustainably (Teece, 2018). Tesla thus represents a complete CASE transformation, combining technological leadership with systemic sustainability.

7.2 Toyota: Hybrid Strategy and Sustainable Innovation

Toyota's long-term vision demonstrates strategic balance between technological continuity and disruptive innovation. While many competitors made abrupt shifts to full electrification, Toyota maintained a diversified portfolio—hybrids, plug-in hybrids, hydrogen fuel-cell vehicles, and battery-electric models—anchored in its *Toyota Environmental Challenge 2050* roadmap (Toyota Motor Corporation, 2023). This multi-pathway approach exemplifies strategic transformation grounded in resilience and environmental responsibility. Toyota's Woven Planet initiative further extends its innovation ecosystem. By integrating autonomous driving research, robotics, and smart-city development, Toyota is advancing beyond vehicle manufacturing toward holistic mobility ecosystems (KPMG, 2023). The Woven City project in Japan serves as a living laboratory for connected and autonomous technologies powered entirely by renewable energy. Such initiatives illustrate Toyota's dynamic capabilities in experimentation, cross-sector collaboration, and adaptive leadership. Moreover, its lean manufacturing heritage, combined with emerging digital capabilities, underlines the firm's strength in continuous improvement (kaizen) and sustainable value creation.

7.3 BMW: Digital Transformation and Mobility as a Service

BMW has undergone significant transformation under its *Strategy NEXT* framework, focusing on digitalization, electrification, and circularity. Its introduction of the "i" sub-brand positioned BMW as a pioneer in electric and connected vehicles, while its ConnectedDrive platform exemplifies integration of AI, telematics, and cloud computing to enhance the driving experience (BMW Group, 2023). BMW's digital architecture enables predictive maintenance, energy optimization, and enhanced driver assistance, embodying the connected and data-centric dimension of CASE mobility. In parallel, BMW has ventured into shared mobility through joint ventures such as *Free Now* and *Share Now*, developed in partnership with Daimler (McKinsey & Company, 2022). These ventures expand BMW's value proposition beyond car ownership toward service-based mobility. The company's emphasis on circular design—through recycling initiatives and sustainable materials—aligns with the circular-economy paradigm. BMW's case demonstrates how established automakers can achieve transformation by coupling digital innovation with environmental stewardship and ecosystem collaboration.

7.4 Waymo: Autonomous Mobility and Platform Ecosystem

Waymo, a subsidiary of Alphabet, represents a leading example of autonomous mobility in action. Emerging from Google's self-driving car project, Waymo has developed an advanced AI-driven system combining LiDAR, radar, and vision sensors. Its operational deployments in Phoenix and San Francisco provide insights into the challenges and opportunities of autonomous mobility commercialization (OECD, 2022). From a business perspective, Waymo's platform model aligns with the innovation-ecosystem theory. The company collaborates with automakers (e.g., Stellantis and Jaguar Land Rover), regulators, and local governments to co-create the conditions for safe, scalable autonomous operations (Roland Berger, 2022). Furthermore, its partnership-driven approach illustrates the open-innovation paradigm—leveraging cross-industry expertise to accelerate learning and market readiness. Waymo's extensive data assets also highlight the resource-based view (RBV) of competitive advantage, as proprietary driving data are essential for continuous AI training and performance refinement. However, challenges remain around ethics, safety, and public trust—underscoring the socio-technical dimensions of CASE transformation.

7.5 NIO: Service Innovation and Circular Business Models

China-based NIO has introduced a disruptive model emphasizing battery flexibility and customer experience. Its *Battery-as-a-Service* (BaaS) model allows users to lease batteries separately from vehicles, significantly lowering purchase costs and enabling modular upgrades (NIO, 2022). This innovation reflects circular-economy and sustainability principles—maximizing resource utilization and reducing waste through battery swapping and lifecycle management. NIO's ecosystem extends beyond vehicles to include lifestyle services, digital communities, and energy infrastructure. Its "Power Swap" stations, capable of automated battery replacement in minutes, illustrate the integration of hardware, software, and service innovation. The firm's digital platform, NIO App, fosters user engagement and brand loyalty by providing mobility services, social interaction, and sustainability insights. This holistic model exemplifies business-model innovation under CASE dynamics, where mobility becomes a personalized and sustainable service experience rather than a product transaction.

7.6 Shared Mobility Platforms: Uber, Lyft, and Beyond

Shared mobility platforms have redefined urban transportation by shifting from ownership to access-based consumption. Companies such as Uber, Lyft, and Bolt embody the shared component of CASE mobility, providing flexible, on-demand transport services integrated with digital payments and navigation technologies. However, the sustainability of these platforms has been questioned regarding their net impact on congestion and emissions. To address this, firms are transitioning toward zero-emission fleets and multimodal transport integration (Accenture, 2023). Uber's *Green Future* initiative, for example, commits to a fully electric global fleet by 2030 in North America and Europe. Partnerships with EV manufacturers and charging providers underscore the collaborative essence of sustainable transformation. These initiatives highlight how shared mobility contributes to the broader decarbonization of transport systems when combined with electrification and renewable integration.

7.7 Synthesis and Lessons Learned

Across these diverse cases, several unifying themes emerge. First, digitalization acts as the connective tissue linking connectivity, autonomy, sharing, and electrification. Second, ecosystem collaboration replaces linear value chains, with partnerships spanning energy, software, and infrastructure sectors. Third, sustainability transitions from peripheral compliance to a core strategic logic shaping innovation and competitiveness. Lastly, leadership agility—manifested in risk-taking, experimentation, and long-term vision—is essential for navigating uncertainty. Together, these case studies demonstrate that successful CASE mobility implementation requires the integration of multiple theoretical perspectives discussed earlier—dynamic capabilities for adaptability, the resource-based view for leveraging digital assets, open innovation for ecosystem engagement, and circular-economy principles for sustainability. They illustrate a global trend toward convergence, where the boundaries between automotive, technology, and energy industries dissolve, giving rise to a new mobility order characterized by connectivity, intelligence, and responsibility.

8. CONCLUSION

The evolution of the global mobility ecosystem under the influence of Connected, Autonomous, Shared, and Electric (CASE) paradigms represents a profound strategic transformation that transcends traditional industry boundaries. Across the preceding analyses, it becomes evident that mobility is no longer defined solely by mechanical engineering or vehicle ownership but by an integrated digital, ecological, and social network driven by innovation, sustainability, and intelligence. The convergence of these forces has compelled organizations, policymakers, and consumers to reconceptualize mobility as a dynamic, service-oriented, and sustainable system. CASE is not merely a technological disruption—it is a systemic shift reshaping economic structures, environmental practices, and urban lifestyles worldwide. As digital technologies mature and climate imperatives intensify, mobility is moving toward an era where data, connectivity, and collaboration will determine long-term competitiveness and societal progress. Yet, this transformation is not without significant challenges. One of the most pressing concerns is the complexity of integration across industries. The shift toward CASE requires deep alignment among automotive manufacturers, software developers, energy providers, telecommunications firms, and regulatory institutions (KPMG, 2023). Managing these cross-sector dependencies poses strategic, technical, and ethical challenges. The mobility sector must evolve from linear supply chains to circular, data-driven ecosystems that ensure interoperability and trust. Data privacy and cybersecurity also emerge as central concerns, as vehicles become increasingly connected and autonomous. Protecting data integrity and user safety while enabling real-time connectivity remains a delicate balance that demands robust governance frameworks and global standards (Deloitte, 2023). Additionally, the electrification of fleets, while essential for decarbonization, raises concerns regarding critical mineral sourcing, energy infrastructure capacity, and recycling of battery components—issues that must be addressed through circular-economy principles and responsible innovation (World Economic Forum, 2023).

A second major challenge lies in the economic and social implications of CASE adoption. The transition to electric and autonomous mobility affects labor markets, supply chains, and urban infrastructure in complex ways. Traditional automotive jobs related to engine manufacturing, maintenance, and logistics may decline, while new opportunities emerge in software engineering, data analytics, and energy management. Governments and educational institutions thus face the urgent task of re-skilling the workforce to meet these changing demands (OECD, 2022). Simultaneously, ensuring equitable access to future mobility solutions remains essential. If autonomous vehicles and smart transport systems are concentrated only in affluent regions, mobility inequality could deepen. Inclusive mobility policies—focusing on affordability, accessibility, and community integration—will be crucial for achieving sustainability that is not only environmental but also social in character (United Nations, 2023). Looking ahead, several future trends are likely to shape the trajectory of the CASE ecosystem. One of the most transformative will be the integration of Artificial Intelligence (AI) and edge computing into mobility systems. Real-time decision-making, predictive maintenance, and personalized travel experiences will increasingly rely on advanced AI models trained on massive mobility datasets. The convergence of 5G and Internet of Things (IoT) infrastructure will further enhance vehicle-to-everything (V2X) communication, enabling seamless interactions among vehicles, pedestrians, and smart cities (McKinsey & Company, 2022). These technologies will redefine safety standards, optimize traffic management, and minimize carbon footprints through intelligent routing and fleet coordination. In parallel, autonomous vehicle development will progress from pilot programs to scalable deployment, though it will require robust ethical frameworks to manage liability, data bias, and human-machine interaction. Another emerging frontier is Mobility-as-a-Service (MaaS), which integrates public and private transport modes through unified digital platforms. MaaS represents the culmination of the shared mobility vision—where individuals subscribe to mobility solutions rather than own vehicles (Accenture, 2023). This trend aligns with younger consumers' shifting preferences toward flexibility, sustainability, and digital convenience. The continued rise of urbanization will further drive the demand for multimodal solutions that blend micromobility, public transit, and autonomous fleets into cohesive, carbon-neutral networks. Moreover, hydrogen fuel-cell technologies and next-generation solid-state batteries are expected to overcome the limitations of current electric vehicles by offering higher energy density, faster charging, and improved lifecycle sustainability (Toyota Motor Corporation, 2023). Integration of renewable energy sources, such as solar-powered charging infrastructure, will create a more resilient and decarbonized mobility ecosystem. However, realizing these future opportunities will require systemic governance and policy innovation. Governments must adapt regulatory frameworks to accommodate emerging technologies while ensuring public safety, privacy, and fairness. Urban planners will play a pivotal role in designing smart, inclusive cities where digital mobility infrastructure coexists with human-centric design. Collaborative innovation—between academia, industry, and government—will be essential for balancing economic growth with environmental and social equity (BMW Group, 2023). Equally, corporate leadership must evolve toward stewardship models that prioritize long-term sustainability over short-term profit. Transparent reporting, ethical AI governance, and stakeholder engagement will determine which organizations thrive in the evolving mobility landscape.

In essence, the journey toward sustainable CASE mobility represents both a technological revolution and a moral imperative. As the world grapples with climate change, energy scarcity, and urban congestion, the mobility sector stands at the intersection of innovation and responsibility. The convergence of digital transformation, circular economy principles, and stakeholder collaboration offers a path toward a cleaner, smarter, and more inclusive transportation future. Yet, achieving this vision will demand not only technological advancement but also a profound cultural shift—one that redefines progress as shared prosperity within planetary boundaries. The mobility ecosystem of tomorrow will thus be measured not merely by speed or efficiency, but by its capacity to connect people, protect the planet, and promote equitable access to opportunity. In closing, the CASE transformation underscores a defining truth of 21st-century innovation: sustainability and intelligence are no longer optional features of mobility—they are its very foundation and future.

REFERENCES

Accenture. (2023). Monetizing digital services in automotive: A wake-up call for automakers to rethink their strategic priorities. https://www.accenture.com/us-en/insights/automotive/monetizing-digital-services-automotive

Barney, J. (1991). Firm resources and sustained competitive advantage. *Journal of Management*, *17*(1), 99–120. DOI: 10.1177/014920639101700108

Bohnsack, R., Kolk, A., & Pinkse, J. (2020). Business models for sustainable technologies: Exploring business model evolution in the case of electric vehicles. *Research Policy*, *49*(2), 103–120. DOI: 10.1016/j.respol.2019.103928

Chesbrough, H. W. (2003). *Open innovation: The new imperative for creating and profiting from technology*. Harvard Business School Press.

Chirumalla, K., Klaff, S., Zako, R., & Sannö, A. (2023). Elevating B2B mobility with sharing autonomous electric vehicles: Exploring prerequisite criteria and innovative business models. *Sustainability (Basel)*, *15*(18), 13757. DOI: 10.3390/su151813757

Deloitte. (2022). The future of mobility: Transforming the automotive industry. https://www.deloitte.com

Deloitte. (2023). Sustainable supply chains and blockchain innovation in the automotive industry. https://www.deloitte.com

Dhanushya, J., Ruba, M., Ponkarthika, M., Soundarya, B., Sakthi Priya, V., Dhanushri, S., & Harikrishnan, S. (2025, April). AI-powered multipurpose agrobot for precision farming: Crop health management, fertilizer recommendation and yield prediction. In *Proceedings of the 2025 International Conference on Recent Advances in Electrical, Electronics, Ubiquitous Communication, and Computational Intelligence (RAEEUCCI)* (pp. 1–6). IEEE.

Elkington, J. (1998). *Cannibals with forks: The triple bottom line of 21st century business*. Capstone.

Group, B. M. W. (2023). BMW Group report 2023. https://www.bmwgroup.com/en/report/2023/index.html

Group, C. A. R. (2021). The value chain is being replaced by the fast-changing automotive C.A.S.E. ecosystem (Connected, Autonomous, Shared, Mobility, Electrified). https://www.cargroup.org/wp-content/uploads/2021/08/Wakefield.pdf

Guanetti, J., Kim, Y., & Borrelli, F. (2018). Control of connected and automated vehicles: State of the art and future challenges. arXiv preprint. https://arxiv.org/abs/1804.03757

Huang, K., Kockelman, K. M., & Gurumurthy, K. M. (2022). An overview of connected, automated, shared, and electric technologies. *Transportation Letters*, *14*(6), 529–540. DOI: 10.1080/19427867.2022.2059703

International Energy Agency (IEA). (2023). CO_2 emissions from fuel combustion 2023 highlights. https://www.iea.org

Kameswaran, A., Nanthakumar, K., & Ruba, M. (2017). High performance precomputation based self-controlled precharge-free content-addressable memory. *Journal of Electronics and Communication Systems*, *2*, 1–9.

KPMG. (2023). Digital transformation in automotive: Driving value in the connected era. https://home.kpmg

Krithika, S., Jayachandran, T., Raghapriya, N. R., Ruba, M., Janani, C., & Sathiyavathi, S. (2025, April). An Effective Denial of Service Texture Features from Malware Grayscale Image. In 2025 3rd International Conference on Artificial Intelligence and Machine Learning Applications Theme: Healthcare and Internet of Things (AIMLA) (pp. 1-7). IEEE. DOI: 10.1109/AIMLA63829.2025.11041501

MarkWide Research. (2023, February). Global connected, autonomous, shared, and electric (CASE) market set for remarkable growth, predicted to reach $1.6 trillion by 2030. Press release.

McKinsey & Company. (2022). The road to seamless mobility: Rethinking ecosystems in the automotive sector. https://www.mckinsey.com

Neckermann Strategic Advisors. (2020). The future mobility value chain: Autonomous mobility, the disrupting piece of the complex puzzle. Medium. https://medium.com/@Neckermann.Strategic/the-future-mobility-value-chain-autonomous-mobility-the-disrupting-piece-of-the-complex-puzzle-c42c0fa0ab4d

Netsol Tech. (2024). CASE (Connected, Autonomous, Shared and Electric): An insight. https://netsoltech.com/blog/case-connected-autonomous-shared-and-electric-an-insight

Puga Garcia, A. (2023, October 23). Towards a new connected, autonomous, shared and electric mobility. MAPFRE Insights. https://www.mapfre.com/en/insights/innovation/connected-autonomous-shared-electric-mobility/

PwC. (2023). Automotive trends 2023: Driving digital and sustainable transformation. https://www.pwc.com

Ruba, M., Ashwin Prabhu, G., Deka, J., Kavitha, D., Sathiya, S., & Pavithra, S. (2026). Empowering learners and educators: The transformative role of conversational AI in personalized education and student support. In Assayed, S. (Ed.), *Enhancing student support and learning through conversational AI* (pp. 163–190). IGI Global Scientific Publishing. DOI: 10.4018/979-8-3373-3316-8.ch007

Ruba, M., Ashwin Prabhu, G., Robinson Joel, M., Vishnu Priya, K., Dharshani, L., & Dharani, R. (2026). Integrating emerging technologies for organizational agility: An interdisciplinary approach to digital transformation. In Roache, D. (Ed.), *High performance leadership for organizational excellence: Innovation, technology, and resilience* (pp. 63–92). IGI Global Scientific Publishing. DOI: 10.4018/979-8-3373-6207-6.ch003

Ruba, M., Girinath, V. P., Ponkarthika, M., Santhini, C., Sakthi Priya, V., & Hemalatha, R. (2023, December). Design and implementation of smart internet-based real-time control of rover with artificial intelligence. In *Proceedings of the 2023 International Conference on Energy, Materials and Communication Engineering (ICEMCE)* (pp. 1–5). IEEE. DOI: 10.1109/ICEMCE2023

Ruba, M., Ponkarthika, M., Parameshwari, V., Girinath, V. P., Soundarya, B., Janani, M., & Santhini, C. (2025, February). Enhancing underwater tunnel safety through smart inspection: An ROV-based system with AI-powered crack detection using Canny edge detection. In Proceedings of the 2025 3rd International Conference on Intelligent Systems, Advanced Computing and Communication (ISACC) (pp. 1067–1072). IEEE. DOI: 10.1109/ISACC.2025

Santhini, C., Ruba, M., Sakthi Priya, V., & Hemalatha, R. (2025). Intercultural communication and assessment practices in the integration of AI with education. In *Generative AI Approaches to Sustainable Development in Higher Education*. IGI Global. DOI: 10.4018/979-8-3693-5623-4.ch010

Shanmugaraja, T., Ruba, M., Priya, V. S., Venkatesh, T., Supriya, M., & Nallamuthu, M. (2023, September). ECG noise elimination using an FPGA version of a modified error normalized LMS adaptive filter. In *AIP Conference Proceedings, 2764(1), 060033*. AIP Publishing LLC. DOI: 10.1063/5.0144367

Tata Elxsi. (2024). Transforming automotive with CASE: Connected, autonomous, shared & electric solutions. https://www.tataelxsi.com/industries/automotive/c-a-s-e

Teece, D. J. (2018). Business models and dynamic capabilities. *Long Range Planning*, *51*(1), 40–49. DOI: 10.1016/j.lrp.2017.06.007

The Tech Outlook. (2023). Connected, autonomous, shared and electric (CASE) market is expected to hit a highest CAGR of 19.2%: Global trends, type, revenue, size, share, manufacturers, key drivers, future scope, supply and demand side analysis. Press release.

Toyota Motor Corporation. (2023). Toyota Environmental Challenge 2050. https://global.toyota.

United Nations. (2022). The sustainable development goals report 2022. https://unstats.un.org/sdgs

World Economic Forum. (2023). Circular cars initiative: Building sustainable mobility ecosystems. https://www.weforum.org

Chapter 2
Emerging Technologies Driving Sustainability in Automotive Supply Networks and Smart Mobility

Mohammad Irfan
https://orcid.org/0000-0002-4956-1170
Chandigarh University Uttar Pradesh, Unnao, India

Saanvi Singhal
https://orcid.org/0009-0002-2676-6816
Christ University, Bangalore, India

S. Dhanabagiyam
https://orcid.org/0000-0002-5268-178X
International School of Management Excellence, Bangalore

Ajay Kumar Yadav
Chandigarh University, Lucknow, India

ABSTRACT

With environmental, social, and governance (ESG) standards at the center of the international business strategies, there is an ever-growing pressure on the automotive industry to be able to guarantee ethical procurement sound carbon footprints and supply chain transparency. This paper examines the lucrative way in which Artificial Intelligence (AI) and Blockchain are soon to transform the sphere of ESG

DOI: 10.4018/979-8-3373-4495-9.ch002

compliance and become the source of end-to-end supply chain visibility. It explores existing issues, namely, fragmentation of data, the lack of its traceability, and the inappropriateness of regulations. The study states the importance of AI to forecast sustainability threats, and the ability of the Blockchain to deliver immutable records of sourcing, logistics, and emissions. It also assesses regulatory bodies, the scale-related problems, and proposes a road map towards incorporation of such technologies, and finally, gives a model of how digital change can be implemented in the creation of responsible, sustainable, and smart networks of automobiles.

INTRODUCTION

The automotive industry has found itself at such a crossroad in a context that reflects environmental concerns on the rise, forces of scarcity, and increased social responsibility. Regulatory bodies, as well as the global consumers have been more demanding of corporate employees to take part in sustainable practice in every aspect of business. Environmental, Social, and Governance (ESG) standards forced businesses where the need to reassemble their position in the sector was not the only necessity but helping the business understand how the use of technology becomes a competitive advantage and the sole chance to ensure the business is firmly in position in the business as well (Krishnan et al., 2024). Such novelty in change as the most important measure to this effect, namely the emergence of new technologies Artificial Intelligence (AI) and Blockchain is also expected to introduce the complete overturn of the kind of automotive supply chains that would become traceable, transparent, and sustainable as such in the long-term perspective (El Jaouhari et al., 2023; Marques et al., 2024).

Sustainability in transportation can no longer be a sidelines objective; it is prime scenery strategic campaign. ESG risks specifically affect the automotive supply chains because their operation is global, they base their economy on coal, and because their enormous dependence on raw materials is being manufactured in the regions commonly referred as socially or environmentally unstable. Legacy closed systems within supply chain management have fallen short regarding the requirement that there had to be an accompanying requirement that the accountability and responsible sourcing existed. This has culminated in the emergence of new paradigms such as intelligent transportation system and smart mobility in the fight against the mentioned inefficiencies of transport and improvement of environmental performance of transport. As stated by Shiva, Vedik, and Manoharan (2024), the following key solution to the future sustainability is brought up in the advancement of the traditional system of mobility when the smart transportation system is created and this system could enhance the sustainability and minimize the carbon-like drawbacks of not only the industrial landscape but also the urban one. Those systems are actively used to integrate efficiency, safety and environmental awareness in the transportation systems of the cities and the industrial supply chains are likely to soon get such digital changeover. To negate the effect of these pressures, companies in the automotive sector must drive further towards the concept of Industry 4.0 in terms of automation of the business process and the changed trends of logistics. Such a problem can be addressed effectively with the help of AI and Blockchain-based digital solutions, managing ESG within a firm, better decision making, and the possibility of accountable sourcing. It is also through its predictive features that AI is increasingly becoming important in forecasting the potential risks to sustainability, optimising the course of these ships to display the most optimal fuel usage, and determining the deficiencies in the activities prior to carrying them out. Instead of it, Blockchain provides the framework of the transparent and permanent record keeping necessary to enable the manufacturer to verify validity of allegations by suppliers, track down the data on the emissions, uphold the compliance with both internal, and external ESG policy. Krishnan et al. (2024) further state that Industry 4.0 technologies have had a considerable effect in promoting the sustainability of supply chains, and the firms have come up with the new, innovative, resilient, and more responsible business models that can adhere to the emerging environmental and social demands (Irfan, Durrah, Chidambaram, & Singh, 2025).

The reasons all lie in the necessity of implementing sustainability in supply chains of the automotive industry promoted by the high carbon regulations, environmental taxation and the global tendency of decarbonization. The Transportation industry is alone that contributes almost a quarter of the global emission and the automotive logistics alone is earning a good pie of this. Marques, Campos, and Cavalcante (2024) note that the connectivity vehicle data context is proving imperative to combine the operations of the transportation systems in a bid to achieve the complex climate goals. Real time analysis and meter reading keep us in the position to make better decisions on how we use the fuel and how to manage idle idle times and emission management which are the most important change agents of sustainability of supply chain. Combined with the immutable data records blockchain indisputably brings, this can allow creation of provable and audible performance histories of ESG performance across whole lines of value: a sort of transparency of records regulators and consumers have never had the ability to access. Nonetheless, in spite of the immense potential of such technologies, application is often limited by a number of systemic obstacles (Dias, Irfan, Galvão, & Morão, 2024). They comprise, but are not limited to, data disaggregation of the supply chain by using legacy technologies, inaccessibility of standard metrics in ESG, and conflict of authority on the international stage. Also, the firms, which establish global networks in the automotive industries are likely to demand the smaller ones producing the supplies and thus they may not possess the technological framework or the material resources to adopt advanced systems. El Jaouhari et al. (2023) suggest an elaborate environmental-grounded model of incorporating IoT in the automotive supply chains to illustrate how the use of digital technologies such as AI and Blockchain could be organized to address these drawbacks. This is an indication that they appreciate the value of a collaborative ecosystem where large manufacturers and smaller suppliers will not only be encouraged and enabled to implement sustainable practices through transparent and technology-enabled frameworks but will also be encouraged to implement sustainable practices (Kismawadi, Irfan, Haron, & Masitah, 2024).

At the same time, the development of the technologies associated with Electric, Connected, Autonomous, and Shared (ECAS) vehicles is also redefining what is expected out of the supply chain. The vehicles are not only ushering in changes in the manner consumers are transported but also driving manufacturers to incorporate intelligence, connectivity, and digital surveillance into the very structure of their manufacturing and distribution supply chain. Vermesan et al. (2021) highlight that AI and data-driven infrastructure can be integrated into the vehicles of ECAS to enable more microlevel ESG tracking, so that stakeholders can track carbon footprints, material provenance and lifecycle effects in real-time. Automaker supply networks have to be 2020-readjusted by using these innovations: they can no longer be seen as linear and shadowy chains but dynamic, transparent ecosystems that are all data-linked.

The combination of smart mobility and intelligent transportation systems also complement the ESG compliant goals. In reference to Elassy et al. (2024), ITS can guarantee the increase of transportation productivity and help to enhance sustainability in a direct way increasing transportation efficiency, alleviating congestions, emissions, and the wastage of the resources. The adoption of these systems in distribution and logistics system on the automotive system can possibly create an impact on the supply chain visibility, short lead time and clean mobility strategies. The platforms on the basis of AI improve the design of the delivery schedules, planning the autonomous maintenance of the fleet, and the regulation of the traffic in real-time and thereby providing the environmentally sustainable logistic system that is circular.

There is also a growth in the governing structure of the supply chain that is being restructured by the enormous degree of integration of the Blockchain technologies. Distributed and hack-proof, Blockchain offers a perfect environment to imbue the aspect of trust in that especially within the environments where the aspects of ESG compliance take precedence. Making reference to ethical sourcing, Blockchain is able to trace the origin of the raw material and even determine the time and work of a working slave which will conform to the established law and results to giving certifications on the environment at every supply chain node. Such technological disclosure does not merely enable a way to have a fit within the regulatory regulations; it can as well help in creating a brand loyalty as well as building consumer confidence. According to Krishnan et al. (2024), businesses which implement Blockchain-enabled traceable solutions can measure their stakeholders confidence level increasing, as well as decrease their risks in operations, and be more competitive in global markets.

EV Production and Emerging Technology Adoption Metrics 2030

The graph displays the seven-years trend of production of electric vehicles (EV) and the implementation of the new technologies Artificial Intelligence (AI) and Blockchain, as well as the influence on the environment expressed in the Greenhouse gases emissions per one vehicle production unit.

During the period, there was a steady increase in EV manufacturing, as the percentages went up to 60 percent, a good sign of the industry moving towards electrification. This fits with the sustainability targets of the world as well as the lobbying by regulators to eliminate internal combustion engines. The fastest change was in the AI adoption rate, which jumped by 77 percentage points to 20 percent, reflecting its potential to be used in manufacturing optimization, autonomous driving system enhancements, predictive maintenance and the optimization of supply chains. Blockchain penetration was another major improvement that went up to 60% against the 15%, indicating that it was being adopted to achieve traceability, smart contract and safe data sharing within automotive chains (Soni, Shaikh, Irfan, & Bijja, 2024).

On the other hand, the amount of CO 2 emitted per vehicle decreased by 5.04 and 2.28 tons respectively and indicated the contributions of a cleaner manufacturing of EVs, AI efficiency, and blockchain-based sourcing and supply chain management (Malcalm, Asiedu, Irfan, & Awudu, 2024).

Such a multi-metric analysis shows that there are evident relations: when the new technologies become embraced, the production of EV increases, and the environmental impact decreases. The paradigm highlights the value of innovation in the automobile industry in not just making automobiles but also playing a significant role in climate objectives by diminishing unit-by-unit emissions. This trend will amplify up to 2030, when the process of digital transformation is expected to speed up.

Figure 1.

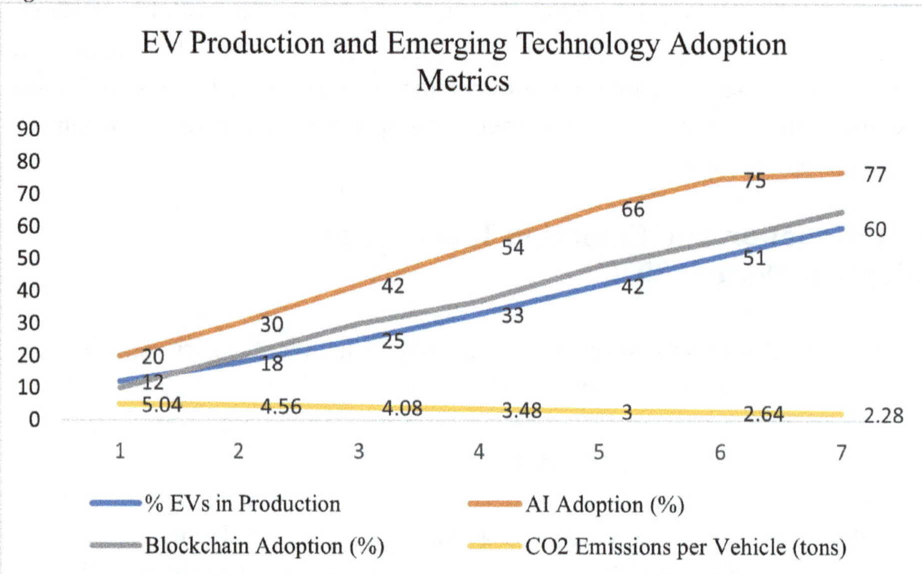

Source: World Economic Forum (2023)

Employing digital technologies has a bright future, but the power of digital will only become the power of ESG transformation to the extent that systems are aligned around organizational, policy, and data systems. To make the transition meaningful, businesses need to transcend the use of the separate digital tools and zero on the development of the integrated platform where the synergistic effect of AI, Blockchain, IoT and data analytics can be realized. This inclusion has to be accompanied by the change in the organization, it is the organization that should train its employees, redesign its operations procedures as well as define new measures to assess its ESG performance. The key determinant in this regard is collaborative industry-wide efforts, government incentives and universal reporting guidelines. A multi-stakeholder model realigns mobility infrastructures along with logistics strategies, with the collaboration of governments, industry leaders, and technology providers. Thus, the sustainable smart transportation must include the multi-stakeholder approach, as Shiva et al. (2024) advise.

Projected Sustainability and Technology Integration Indicators for 2047

The named chart is a prospective assessment of the degree to which the automobile and manufacturing industries can be sustainable in regard to sustainable goals and innovative technologies. It shows that by 2047, the amount of vehicles that will be electric will be 87 percent, which shows an enormous transition among industries in the entire world, by ending the reliance on fossil fuels. The projection is a summation of all the regulatory push, climatic pledges, and consumer pressure starting to cleaner solutions to mobility.

The next big change is the one in the complete implementation of Artificial Intelligence (AI) in supply chains, and it is predicted to be implemented by 98 percent. It indicates that by 2047, AI will support the industrial processes by providing real time data treatment, intelligent logistics, proactive recovery, and production planning with as little energy consumption as possible. New Supplier traceability via blockchain is expected to be at 76%, showing the importance of the blockchain technology to develop tamper-proof and secure records that can address the ethical sourcing, labor, and environment-related carbon footprint tracking in the global supply chains.

Nevertheless, with the technological breakthrough, not all firms can expect to reach net-zero emissions, which shows the complexity and challenge of a complete decarbonization. The source of this shortage can either be technological, financial or infrastructural where barriers still exist in some industries. The sustainable materials used in production in the meantime are estimated at 82%, which is a good indication of improvement but still does not cover all the aspects, hinting at continued usage of finite or few resources.

In general, the data is an indicator of a future in which technology becomes the driver of sustainability, and electric mobility, AI, and blockchain are leaders of the change. Although the advancement is encouraging, the current disparities in net-zero realization and material sustainability mean more profound innovation, policy coordination, and business obligation are considered essential to fully execute low-carbon, circular economy by 2047.

Figure 2. Sustainability and technology integration indicators for 2047

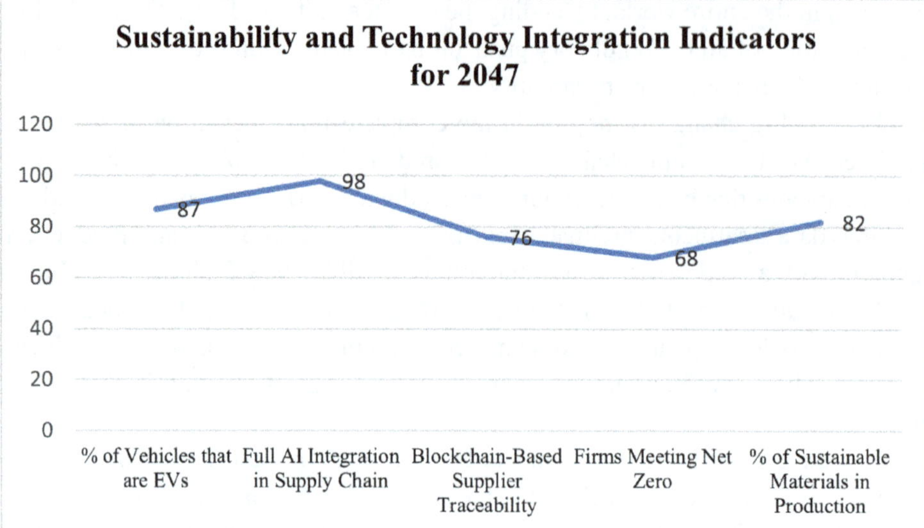

Source: *Future Mobility and Sustainability Trends Report (2047 Projections), compiled from World Economic Forum, IEA, and UNEP scenario models*

The following paper aims at exploring how emerging technologies can be used to power ESG compliance and sustainable mobility in the automotive sector. The paper seeks to provide a detailed analysis of AI and Blockchain applications with the help of practical examples in the industry and model frameworks to address the concept of digital-driven sustainability in a holistic manner. The study examines the way in which these technologies can solve particular issues of the ESG, including ethical sourcing, emissions monitoring, and regulatory compliance, as well as what operational and policy-level enablers are required to achieve such a solution. Finally, the paper is meant to provide a strategic guide to companies in the automotive industry to traverse through the new reality of sustainability, regulatory burden and digitalization to align supply chain processes to the global ambitions of climate resiliency and sustainable global industry.

LITERATURE REVIEW

The move to eco-friendly transportation has seen the car industry become the epicentre of worldwide decarbonisation and innovation policies. A lot of literature has been developed in the recent decade with regard to the relevance of digital transformation and new technology in the building environmentally responsible supply chain ecosystems Krishnan et al., 2024; Masoumi et al., 2019). Researchers have studied how intelligent systems, blockchain, and AI combination can make systems transparent, optimistic, and aligned on environmental, social, and governance (ESG) strategies Aggarwal, 2024; El Jaouhari et al., 2023.

Smart transportation The key element of sustainable city development is smart transportation, which involves the connection of old mobility patterns with new levels of digitalization, minimizing the number of emissions, traffic jams, and misgivings. Shiva et al. (2024) note that the urban logistics can be greatly enhanced through the introduction of intelligent transport system and contribute to the green and safety goals. These systems are based on real-time data collection combined with the ability to make optimization and faster decisions by utilizing artificial intelligence. In the same spirit, Elassy et al. (2024) point to the benefits of constructing intelligent transportation systems in terms of its potential to make cities more interconnected and sustainable, opening the door to greener and efficient supply chains in automotive logistics.

The concept of Industry 4.0 has turned out to be inspirational in achieving supply chain sustainability particularly in the automotive industry. (Krishnan et al. (2024) provide an exploratory research describing the benefits of business model and sustainability innovation that can be found through implementing cyber-physical systems, big data, and the use of digital platforms. Their conclusions underline the importance of predictive analytics and digital interconnectivity in minimizing the waste, better resource distribution, and in greater accountability of the suppliers. The same is discussed by 9El Jaouhari et al. (2023) who introduce a framework that takes advantage of IoT to build automotive supply chains traceable against environmental sustainability. Such models support a real-time control of emissions, energy consumption, and materials, which are in line with expectations regarding ESG compliance.

With the world moving towards clean power and low-carbon technology, electric vehicles (EVs) and intelligent mobility formats are catching up. According to Vermesan et al. (2021), Electric, Connected, Autonomous, and Shared (ECAS) automobiles are transforming in the transportation sector, and to some degree, the technologies involved entail integrating AI-based intelligence as well as decentralized data control structure with transport architecture. Not only does this improve user experience, it also permits a smooth tracking of indicators of environmental performance throughout the life of the product. Simultaneously, Kashem et al. (2024) address the issue of renewable energy adoption in EVs and future flying vehicles as one of the drivers of intelligent mobility systems. Besides the revolutionary definition of mobility, these innovations put manufacturers to the test, in terms of designing sustainable supply chains, which are conducive to new energy capabilities and production methods.

Other research fields include utilization of connected vehicle data to manage carbon. Marques et al. (2024) stress the importance of vehicle telematics and connected data systems in the practice of transportation-related emissions management. It is important to note in their study that near real-time data about the vehicle operations including idling, route followed, and fuel use can be used to create real-life emissions profiling and suggest sustainable actions on these vehicles. This helps to improve on ESG reporting and provides companies with the benefit of being able to proactively maintain and minimise their environmental footprints at the supply chain.

Though the developments in vehicle technologies are a promise, the deeper issue is the challenge of integrating the sustainability approach throughout the entire levels of the supply chain. Masoumi et al. (2019) perform a systematic review of sustainable supply chain management in the automotive industry with regard to the long-term planning, green procurement, and life cycle analysis. Their studies give some understanding of the history of the supply chain sustainability development since it should be supposed that digital tools should be combined with policy and cultural shifts to achieve their full potential. Likewise, Anastasiadou (2021) presents a decision-aiding methodology according to which new vehicle technologies are ranked on the basis of their impact on the sustainability, which further strengthens the idea of applying multi-criteria analysis to automotive planning.

Through automation and connectivity, the building of business models in digital mobility is changing the way in which auto firms add value. Perez-Moure et al. (2023) focus on the ways companies like Volkswagen and Toyota can handle the pressure of sustainability with the ecosystem of their connected vehicles and service-based modi operandi. In their analytical process, they indicate how data-based websites and applications can enhance the performance of vehicles by decreasing the time when vehicles are not operating, improving fuel consumption, and ensuring that consumers are engaged due to eco-friendly functions. The digital transformations are necessary in order to continue supporting competitive capabilities of companies at the same time complying with ESG requirements.

Blockchain has received a lot of traction as a remedy to the transparency supply chain. Aggarwal (2024) provides an extensive overview of the potential of blockchain to improve traceability, accountability of supplier and ethical sourcing. Blockchain can certify a company with its adherence to labor laws, emissions limits, and sustainability certifications by establishing tamper-proof digital records. It is especially crucial in the automotive industry, where global supply networks can be characterized by numerous levels of suppliers that work in different regulatory environments. Blockchain, therefore, turns into trust-enabling infrastructure to stakeholders in geographical boundaries.

The climate dynamics in transportation are also being transformed by disruptive technologies like the biofuels or autonomous vehicle. Chester (2015) also examines the possible role of these innovations and on-demand mobility services in reducing the environmental impact. The paper points out to the second order effects of technological disruption like reduction in vehicle ownership rates and energy consumption. These revelations favor a wider ecosystem perspective of sustainability in that transportation efficiency and behavior change are inseparable with digital transformation.

The supply chains of electric vehicles have been reviewed strictly on their green basis provision of environmental impacts. The article by Esiri et al. (2023) performs a life cycle analysis of EV supply chains and pinpoints the spots with high impact concerning material sourcing, production, and disposal. Results of their study emphasize how it is vital to incorporate environmental values into all processes of production and distribution. AI-based forecasts, and blockchain used to validate them, become the feasible way out to these sustainability gaps.

New ideas like Transportation 5.0 also contribute to this movement and promote the comprehensive combination of intelligent technologies and renewable energy. Shamsuddoha et al. (2024) define the potential of this model of transformation, underlining automated dimensions, clean electricity, and system-level coordination. Their vision revolves around the adaptive, data-profiled automotive ecosystems with the environment and consumer responsiveness.

Lastly, logistics are experiencing logistical changes leading to a creation of smarter systems. One of the ways in which literature and real-life case studies converge in issues of climate change and digital innovation (Bueno-Pascual, 2024) is through transportation and logistics. The trends of establishing urban consolidation hubs, inventory management based on artificial intelligence and predictable maintenance are pointed out as both a cost and a sustainability driver. These macro changes are necessary to the movement out of separate supply chains to cohesive, ESG compatible networks.

RESEARCH METHODOLOGY

The paper embraces a mixed-methods research method to discuss the synchronous relationship between Artificial Intelligence (AI), blockchain technology, and sustainability in automotive supply chains. The reason of taking this is because of the fact that the topic is multidimensional thus including both quantifiable operation effects to it and qualitative arrangements in governance and transparency. The research should also include a comprehensive and informative presentation on the role digital innovations are playing in improving ESG compliance and transparency in supply chains in the automotive industry by merging both qualitative and quantitative data.

The research methodology of the work is based on designing descriptive-exploratory one, leaving space to reflect upon the future trends and the practices. The exploratory element will be especially important since AI- and Blockchain-based technologies have been relatively new when it comes to ESG-oriented supply chains. Although these technologies are the focus of more and more conversations they are not widely utilized in most companies, and it is therefore critical to research how front-runners are using these technologies to become more productive and transparent. Descriptive part permits categorization and comparison of use cases, performance results, and strategic models which are currently in use by automotive firms globally.

The main body of this research is secondary data. The process of the systematic review of academic articles, white papers, corporate sustainability reports, the case studies in the industry, and government publications has been done to extract the information as to how AI and Blockchain are commonly applied in the automotive industry. The study must be of scholarly rigorous standard which is attained by involving peer reviewed academic research, yet the existing industry reports must be also involved so that the study be applicable in the actual world. Not only can such dual source model satisfy the gap that exists between theory and practice but it is as well essential when one has to deal with changing technological landscapes at a very fast rate. They consisted of databases, such as Scopus, Google Scholar, IEEE xplore, and Sciencedirect which are to be searched as thoroughly possible with the keywords such as AI in automotive supply chains, Blockchain ESG compliance, sustainable automotive logistics, and digital supply chain traceability.

These were thematic analyzed and sieved and coded in such a way that the new literature could be sieved and marked out substantially with common trends, the most critical success factors, and the limitation of technology. These qualitative analyses helped to come up with the baseline knowledge of the use of AI algorithms in routing optimization, reduction of wastefulness, and forecasting of ESG threats, including predictive analytics, machine learning, and automation. Meanwhile, the application of Blockchain has been studied with regard to traceability, fraud verification, regulation checking controls and ethical sourcing that is vital in achieving the supply chain transparency.

To supplement the research findings, some case studies were sourced among the developed automobile corporations and the global sustainability initiatives. These case studies, among others (McKinsey reports, WEF publications, corporate disclosures e.g. Toyota, Volkswagen and Tesla), show how leaders in the industry are integrating the emerging technologies into the value chains. All the cases were considering in a defined framework: problem identification, technology application analysis, outcome evaluation, and ESG applicability. This can be used to make cross-case comparisons and at the same time keep the depth essential to comprehend firm level contexts. As an example, the implementation of Blockchain in sourcing tracking or predictive maintenance with the help of AI of Tesla or BMW CI respectively were analyzed to be able to see the technical implementation and ethical results.

To supplement the secondary data, transcript of industry webinars and panel discussions by industry experts and their comments was also included. These sources provide relevant remarks on real time challenges and opportunities on the observations of the practicing persons, although it is not formal interviews. The authority will also be given to the qualitative part of the research through the interviews, therefore complementing the intellectual knowledge with practical dilemma. Thematic coding of statements presented by supply chain managers, tech providers, and sustainability consultants was conducted to identify attitudes to use of technology, hindrances to the integration and future prospect of such products. The quantitative approach to the methodology is founded on the implementation of the comparison between the indicators of ESG performance that preceded technology implementation and the one that occurred when technology had already been introduced to regions where such data could be retrieved. Some of the measures that were taken into consideration were carbon emissions per unit of a product, the amount of energy consumed at different levels of supply distribution, time to carry out an audit as well as the amount of materials that were ethically sourced. This is the figure which was downloaded in the reports on sustainability published on the web and was compared against the average in the industry. Although there were shortcomings in regard to the range of data that would be captured there was the trend analysis which was still being employed in order to end up carrying out an indicative analysis of the level of measurable aspects of the technology. This was not to find only the correlations but to chart out the facilitating elements, in the use and in combination of AI and Blockchain, to drive up ESG scorecards.

It is also here the contextual sensitivity is also highlighted greatly in this research. Since the supply chain practices that are considered have been identified as those that are geography specific owing to the variations in various regulations, employment rules and preparedness of any infrastructural investment, the methodology that would be involved would be a comparative one. One of the examples was the supply chain technology being introduced to the European Union and the emerging markets which have the knowledge of the voluntary disclosures (understanding of ESGs). This comparative layer contributes to the understanding of the nature of policy environments and the effects they have on using the emerging technologies and their success in sustainability supply chain. Besides that, this study is also aware of the fact that it may introduce bias and weakness on the use of secondary data. Corporate reports and case studies can also be used as they possess the valuable information, however, it might not be reporting-biased, particularly in terms of self-reported ESG success. To address this, triangulation was used- checking the validity of the claims made by using different sources e.g. academic reviews, evaluation of claims leveled by third parties (e.g. CDP and GRI reports) and independent analysis of a particular case. In the situation of conflict, the least outlying piece of data computed was considered so as to guard the integrity of findings.

The methodology issues on ethics were as well addressed. No significant ethical concerns were related to human subject proactive involvement or primary information gathering that would be obtained due to the interviews, thus the ethical approval was not expected. However, the due diligence regarding the respect of intellectual property rights and proper referencing as a result of citing the sources was preserved as well as the transparency concerning the interpretation of data itself. Also the study failed to make absolutes where it was evident that the research was pointing in the matter as opposed to ascertaining it. This was especially to the subjects in new areas that had yet to be deprived of empirical credibility. The ultimate outcome of this methodology is the effort to propose a framework of the AI and Blockchain integration in the sustainable supply chains of automotive industry. It proposes seven key steps that are listed in this literature and evidence-based conceptual model: the selection of technologies, engagement with the stakeholders, cross-alignment on ESG, plan of implementation and performance tracking. The framework is both scholarly and practice-oriented when it comes to assisting automobile companies to begin the process of digital sustainability transition.

In conclusion, it can be stated that the research methodology used in this study is constructed to present the best balance between the theory and practice. The combination of qualitative exploration and quantitative knowledge, with the case studies being placed under the premises of national policies and industry, the method ensures the multi-dimensional depth of study. With the automotive supply chains rapidly being confronted with the two intertwining forces of digital disruption and climate responsibility, the given study becomes a present and practical addition to the research endeavor of sustainable operation and supply chain innovation.

Flowcharts

Figure 3. Digital Technologies Driving Sustainable Supply Chain Transformation

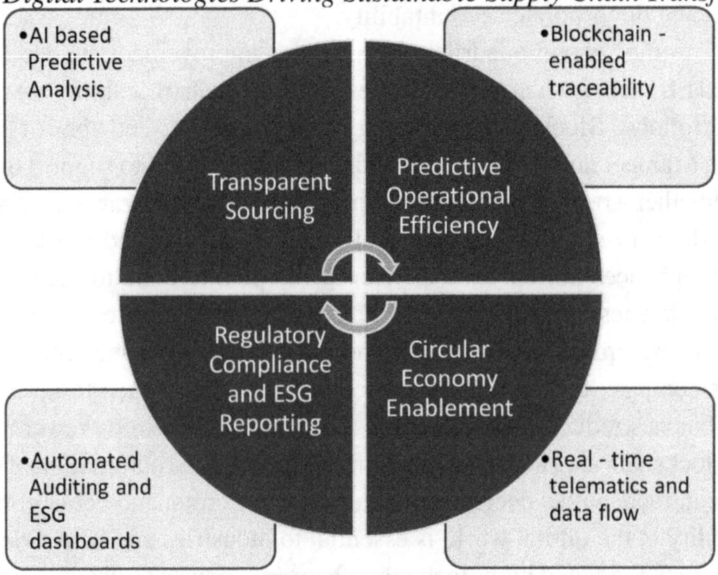

As shown in Figure 3, Author Estimated, the following flowchart reveals an overarching structure of how the digital technologies are transforming the sustainable supply chain practices. The illustration illustrates a circular map that is categorized into four major blocks with Transparent Sourcing, Predictive Operational Efficiency, Circular Economy Enablement, and Regulatory Compliance and ESG Reporting. These are the main areas, each of which is facilitated by the innovative digital technologies, which allow smarter, more effective, and more environmentally-friendly operations. This networked structure is an evolution of the normal linear supply chains to more vibrant, information-based and resilient types which incorporate the needs of any contemporary business and environmental practices.

Transparent Sourcing is considered in the first quadrant, which underlines the significance of sourcing raw materials and components traceability and visibility. Predictive analysis is significant with the help of AI in this respect. Through artificial intelligence and machine learning, organizations can study and examine large bodies of data and forecast the risks of suppliers, ideal sourcing locations, and changes in the market. With this analytical capacity, the companies have the option of making informed choices consistent with their sustainability ambitions. Predictive analysis is also about aiding in the forecast of demand and prices and assisting in morally upright sourcing by determining possible problems of labor or environment violations in the supply base. Fueled by the rising need of consumers and regulatory organizations to see more transparency in their sourcing efforts, AI-assisted intelligence contributes to developing a more responsible and agile sourcing strategy. This builds credibility towards its stakeholders and aids in the greater concept of corporate accountability.

Shifting to the second quadrant, the inherent improvements under Predictive Operational Efficiency when applied in view of blockchain traceability are extremely clarified as follows. Blockchain technology offers an end-to-end view of the supply chain using a tamper and decentralized ledger making it ideal to support blockchain software. In other words all transactions and the process of the raw materials supply to the end delivery of goods can be safely logged and checked in real time. The traceability enhances not only the efficiency in operation due to the reduction of errors, redundancies and fraud but also responsiveness in case of disruptions or recalls. As an example, when a company produces a faulty product, blockchain will enable this company to find out the source of the product much faster and isolate particular batches reducing the company its reputation, harming fewer people. In addition, blockchain aids in the verification of suppliers and automation of contracts via smart contracts with a decrease in administrative costs, and better compliance. Such visibility in the digital world is essential to industries and categories such as pharmaceutical, food, and high- tech manufacturing industries where accuracy and quality are important elements. Finally, blockchain-based system generates confidence, decrease expenses and promote operational perfection due to its unchangeable and announcing infrastructure.

Circular Economy Enablement provides a third quadrant, which concentrates on the most effective use of resources and reducing waste, and is beneficial to a regenerative production and consumption model. This transition occurs on the flow of real-time telematics and data. Telematics solutions that combine on-board diagnostics and GPS tracking, give a real-time view of fleet operations, how assets are used and logistics. The systems help organizations minimize the amount of fuel used, route delivery vehicles more efficiently, and correct equipment before it breaks down to lengthen equipment life. In addition to logistics, real-time information flow through the entire chain of supply provides the opportunity to monitor material consumption, the rate of recycling and logistics of returning. These processes are key elements of the circular economy. On an example, the manufacturers can trace how the recyclable or biodegradable materials are used, control reverse logistics of returned and used products, and to create a product which could be dismantled and be remade. IoT sensors and edge computing even allow an organization to automate the process of sorting, reprocessing, and redistributing materials. Real-time data can enable stakeholders to transform to a circular kind of economy, by replacing the current ways of a take-make-dispose economy, with one in which value is constantly derived out of resources, and wastes are kept as minimal as possible. This is not only in harmony with the regulations on the environment, but also introduces new channels of revenue due to the reuse and resale of the products and materials.

Regulatory Compliance and ESG (Environmental, Social, and Governance) Reporting is the fourth quadrant that is increasingly becoming more and more valuable as the world is entering the era in which regulatory requirements and investor pressure grow more demanding than ever. The comrade of this segment is an automated auditing process and the ESG dashboard, which significantly streamline the compliance and reporting part. The further and further the regulators put their pressure, the more the investors and the consumers will want firms to provide accurate and detailed report of their impact on the environment and their labor practices, the types of governance, the extent of their activities that can be considered as sustainable, etc. The auditing tool based on AI and data analysis enables the supply chains to actively monitor compliance within their operations with respect to an existing framework, such as the ISO standards, GRI guidelines, and state guidelines. This is a pro-active monitoring whereby the organizations can get to know the risk of not being compliant early enough and correct the situation before it explodes out of proportions. ESG dashboards will gather information captured by various systems, including supply chain management systems, HR databases and environmental monitoring systems, and present it as the single point of visualisation. This not only enhances decision making, but also enhances transparency to the stakeholders. Emissions levels and reports, diversity level and supply chain ethics reports are available in a real time manner hence a company will be able to rate the long term staying power and ethical standing of a company. Automation also reduces manual effort, reduces the cost involved in carrying out the auditing process, and, as such, not only it reduces the number of mistakes but is used to generate more accuracy and validity in the reporting process of ESG. Automated systems are getting to be increasingly significant because of the liquidity and precision required to handle emerging customer and investment requirements in the face of ESG as an aspect of investment and customer decision-making. The synergetic connection between technology, strategy and sustainability is the most basic of the loop of this kind

Bidirectional arrows in the middle of the figure help to stress the interdependence and the flow of data, value, and accountability of all the four segments. This model promotes the idea of a complex approach to the components of a supply chain and not a stand-alone system since the improvement of one area transforms to improve the others. An example of this is the blockchain traceability: an exercise that does not only enhance the operational efficiency but also directly contributes to the ESG reporting via providing unchangeable records of sourcing and logistics data. Likewise, AI-based sourcing practices may support the implementation of the circular economy initiatives in terms of prioritization of suppliers with the experience in working with recycled materials or close-loop supply systems.

Finally, in the closing part, Figure 1 offers a future-proof vision of the supply chain i.e. one that is intelligent, transparent, sustainable and resilient. AI, blockchain, IoT, automation, and other digital technologies are less supplementary features that can be integrated but the key enablers of sustainable transformation of supply chains. Integration of these technologies into the sourcing, operations, circular processes, and compliance operations of an organization will not only allow them to ensure that the environmental footprint is minimized but also make it an innovation-driven process allowing them to build better relationships with their stakeholders and accomplish long-term business success. This macro perspective reveals that the digital integration should be used as a means to transform the sustainability regulatory burden into a value-creating strategic opportunity leading to an ethical, optimized, and ecologically friendly global supply chain.

Figure 4. Digital Transformation Pathway Toward Sustainable Urban Mobility

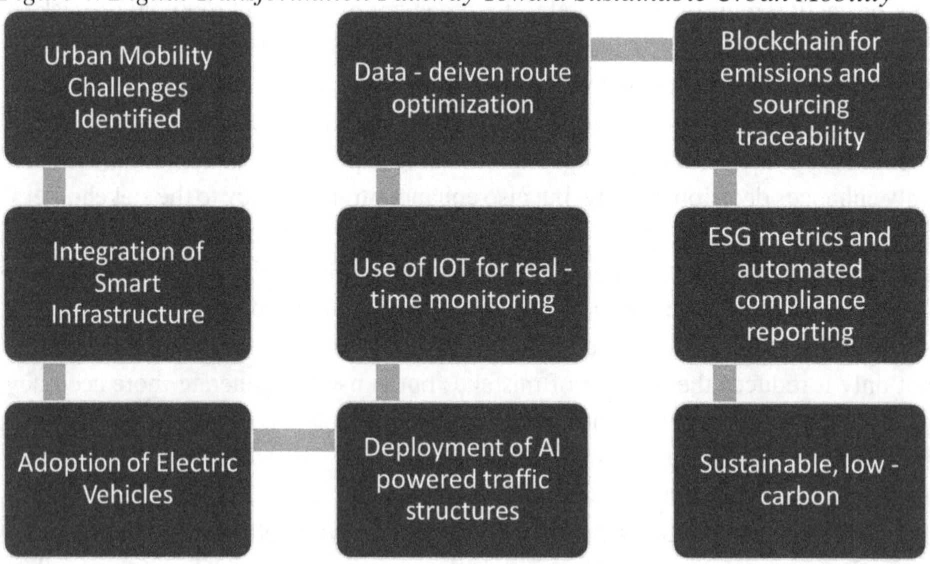

Figure 3, Author's Estimated, shows, there is a logical pattern of initiatives and technologies, which, when used together, can find the means of transforming existing, sustainable, low-carbon urban mobility systems. The presented diagram is the strategic approach that describes opportunities to address the increasing issues of modern cities regarding traffic jams, pollution, energy consumption, and ineffectiveness of the public transportation system. It is graphically depicted as a step- ladder starting with identification of the problems of urban mobility and ending with actual achievement of the low carbon and sustainable transportation ecosystem.

It is possible to support this path with introducing the latest digital technologies, environment management tools, and infrastructure advancements.

Find the Urban Mobility Challenges is identified as the initial step in the flowchart. It entails identification and discussion on the major problems bedeviling the urban transport system, including traffic jams, air pollution, poor structure, and overdependence on fossil fuels powered vehicles. The urban planning authorities and policy makers determine such issues by the opinion or participation of the stakeholders, mobility data examination, and environment-related studies. Once such obstacles are normal, built-up spaces can have a guideline to operate sustainable transportation: Unassisted, Multidimensional, and low-environmental impact.

Based on this premise, the second phase of the process introduces Integration of Smart Infrastructure that acts as a driver of digital and sustainable transformation of the city transport. Smart infrastructure also includes digital traffic lights, networked buses and trams, traffic flow sensors (real-time) and smart road surfaces. They are such elements to acquire and transfer data that makes the traffic dynamic and the infrastructure to be maintained. With this integration, the digital back bone can be used further to create more innovation such as Internet of Things (IoT) and deployment of artificial intelligence (AI). Using the physical infrastructure of the city along with the intelligence added to it, the municipalities will be able to optimize the traffic flows, reduce the downtimes of the physical infrastructure and the life of the commuter.

Meanwhile, EVs Adoption also turns out to be one of the most significant pinnacles of such a change. On the one hand, compared to vehicles with the internal combustion system, EVs are only one of the solutions to their chief issue of air pollution in urban areas and emission of carbon. But to adopt it massively, the EVs must be marketed through charging facilities as well as policies and integrations of EVs into the available systems of a shared and public transport system. Most advanced cities offer free tax credits, low tolls, and priority lanes in a bid to hasten the level of EV use. The shift to EVs will help to deliver the long-term plan of zero-emissions transport system as long as the electricity sources are renewable or of low emissions. Data-driven Route Optimization, in its turn, is the main of the supposed abilities that metamorphose into the possibility with the help of the concept of Digital transformation along with the Infrastructure and EV integration. With the assistance of the refined analytics, GPS data and trends of the traffic history, cities and logistic companies can optimize the authentic path of the public and the private transport. Such optimisation minimises the time spent travelling; the consumption of gas or energy and the satisfaction of the user is increased. The re-routing of the real time data can be dynamic such that it can be able to meet the traffic demands, highway blockages or even weather delays in order to ensure minimal impact on the fullness of the mobility economy. This also means that the optimized commercial supply chains not only ease off the congestion and emissions of the freight in the urban regions which incidentally is not covered in some scenarios to be an issue contributing to poor air quality in urban regions.

The second level is the Utilization of IoT in the Real-time Monitoring, the basis of which is the degree of smart infrastructure and data analytics. The traffic signs, the road and the cars are still being checked with sensors that collect and relay data to place of centralization. The real-time feedback loop helps the cities to be informed of the traffic density, the traffic activity, prevalence of air pollution, and the conditions of the roads. Based on this data, it will be possible to manage dynamic changes of traffic flows, actively maintain infrastructure and control environmental conditions. The capability to develop a flexible and smooth urban mobility environment depends on responsiveness of systems of IoT.

Besides IoT, it is the Deployment of AI-powered Traffic Structures, which is the leap of automatized and predictive planning of municipal transport networks. AI algorithms can analyse the large amounts of data gathered by the IoT devices to e.g. forecast traffic, optimise signalling and decongest hotspots, or even foresee when devices must be service. These AI systems will augment the operations of the traffic and allocation of resources and such systems will also make the urban traffic more sustainable, easier and safer. In addition to regulating the traffic, another application of the AI in the decision-making process might be the fleet optimization, planning the work of the public transport as well as studying the habits of the user of the transport and making the operations even more efficient and the environment-friendly.

Parallel strand Blockchain in the Emissions and Sourcing Traceability provides transparent safe way of tracing of the emission and assurance of transport component and structures sustainability. There are possibilities to record carbon traceability of cars, supply chain emissions of car parts or source of energy to charge an EV in blockchain which provides append-only books. Consumers with such traceability would be in a position of making not only environmentally but ethically sound choices which would assist them to be on the good side of the regulatory demands. Further, blockchain also verifies carbon at offsets and makes it possible to publish open reporting to environmental activities of the mobility ecosystem on the aggregate level. The integration of the ESG Metrics and Automated Compliance Reporting and assessment of the blockchain will assist organizations and municipalities to align their operations with global standards of environmental, social and governance aspects. Automation observes, gauges and documents emissions, resources utilization, labor relations and governments policy. These are pointers that will help in facilitating decision making, this will give confidence to stakeholders and it will increase accountability. The policymaker will be able to keep track of the development of the mobility work and become acquainted with what should be more heavily invested in or what initiatives should be taken with the assistance of dashboards and analysis tools. It also minimizes administration and human error and thereby reporting compliance becomes the norm in the sustainability reporting. All the above will culminate in the end deliverable as depicted in the figure which is the development of Sustainable Low-carbon Urban Mobility System. Such end game scenario can be termed as how we have an energy consumption powered by clean energy in transport, full automation of smart transport and transportation powered by real time data. It is the mobile environment that offers the environment, efficacy and social justice. Green urban mobility minimizes greenhouse gas emissions, offsets the traffic jams, ensures air quality and increases the quality of life of the city populations. Notably, it will assist in achieving international climate targets as spelt in the Paris. The international climate accord and SDGs of the United Nations. A flowchart as shown in Figure 2 shows that sustainable urban mobility is not a simple endeavour but a multifaceted endeavour with intertwining factors surrounding technology, regulatory, and behavioural modification. It needs an effort of collaboration of city planners, government agencies, technology providers, transport companies and the general population. The sequence illustrated in the figure stresses that the digital technologies of the AI, IoT, blockchain, and data analytics are not the lonely ones, these are the inseparable parts of a more comprehensive change plan. The specified technologies can help cities to address the deeply rooted mobility problems and develop a resilient, inclusive and future-proof transport sector when combined with the strategic implementation.

Finally, Figure 2 gives a detailed but imaginative view of digital transformation of urban transportation. With a coordinated approach to resolve the structural problems, a long-term investment into creating smart infrastructure, clean technologies and making decisions based on data, cities will be able to shift to the intelligent mobility network, leaving behind the congested transport and polluting infrastructure. This change is not merely necessary to protect the environment but also key to economic growth, healthy population and social welfare. It is a dynamic and futuristic vision that is summarized by the visual representation of the flowchart and directs the stakeholders towards the vision of a cleaner, smarter, and fairer urban future.

CASE STUDIES

Case Study 1: BMW Group

The BMW Group is an organization that globally carries out the objectives of incorporating developing innovations in its automotive supply area in order to balance out its sustainability. The core of the strategy of BMW is the so-called PartChain platform within the supply chain-based on blockchain that makes the components and materials traceable. The invention will help the BMW to trace their source of parts they are using and will use hence ensuring that the suppliers are friendly to the environment and to the society as well especially the rare materials such as cobalt and lithium which will be used to manufacture the electric vehicle (EV) batteries. Blockchain can arrange an erasable audit trail of provide chain data as a means of delivering additional transparency within the supply chain and lowering risk of fraud or immoral sourcing opportunities. PartChain will allow BMW to see the movement of a part in the remainder of part (extraction of the raw materials, processing using the assistance of various devices and assembly in the end) and will F by easier to recognize what effect a step has on environment. This would enhance its adherence to the sustainability objectives and would gauge it well in the environmental, social and governance (ESG) reporting. BMW also uses the Internet of Things (IoT) not only on plants but also on its supply chain, in addition to blockchain. Equipment and vehicle sensors gather real time data, regarding conditional data that include temperature, location and performance data. Such information is used under the predictive maintenance systems, thereby eliminating equipment failure, downtimes and enhance energy savings on a batch per the plant. Artificial intelligence (AI) does not also leave out the sustainability of BMW. One can mention its system named the Proactive Care which has combined AI and better prospects of monitoring the conditions depending on the information about using the vehicles. This not only helps in serving customers but also helps the car parts to have a longer life thus preventing wastages of materials and environmental impacts. The transition to the circular economy is due to the digital skeleton of BMW as well. The company is coming up with mechanisms through which an individual can employ in tracing down things like aluminum and parts of batteries and re-gaining use of the same. Blockchain is capable of assisting in this verification process and so there will be no such need to use virgin materials as these wastes can be retrieved back and introduced back into its production chain.

Besides this, BMW has in place real time dash boards and analytical tools to monitor carbon foot, energy expenditure and wastes in their operations. The tools make sure that the company is producing and achieving the targets on the sustainability as they are pragmatical in terms of providing feedback. Representatively, BMW has made an initiative of reducing CO 2 emissions within its entire value chain and monitors the same with help of the web-based tools. Also, MW is involved in the smart mobility programs such as test riding of the car sharing EVs, and automated driving within the city. This is done in an attempt to decongest such areas, and it also strives to have the proper transport ways that will be encouraged. In short, the BMW remodels Lake manufacturing and supply chain process, and it aims to integrate block chain, IoT, and AI solutions to help it to develop the industry that will be less polluting to the earth. BMW has the chance to achieve the maximum environmental performance, and its operating efficiency considering its open and analytics-driven ecosystem that would enable the company to be a forerunner of digital advances in terms of sustainable mobility globally.

Case Study 2: BYD

BYD (Build Your Dreams) is a Chinese company, a leader in the electric vehicles (EVs) market, and a leader in the green car manufacturing. BYD has a vertically integrated architectural model, unlike most carmakers who outsource most of their parts, but manufacturers most of their parts mainly batteries, semiconductor, motors etc,. This organization allows BYD to have a tight control of its supply chain and also easy incorporation of sustainability in all the stages. One of the most outstanding innovations is the Blade Battery of the BYD that is safer, longer-lasting, and more efficient to use in comparison with influential lithium-ion batteries. The IoT sensors on these batteries record the data in real-time on temperature, charge cycles, and their operational performance. The analysis of the data via AI is performed so that to optimise charging behaviour, extend the battery lifetime and minimise the energy wastage. Some products BYD manufactures in its plants are smart factory systems, which integrate IoT and AI to streamline production lines and conduct quality control in real-time. This minimizes defects wastage and use of energy latency. Such as they have sensors that scan and tell how the production is going on and adjust to the setting of the machine using the least source of electricity and least wastage of the material with the help of an AI-integrated system. It is also said to power majority of the factories using renewable energy or in other words the solar panels produced on its own energy department. Such utilisation of clean energy is very beneficial in reducing the emissions of cars during the manufacturing of automobiles and this could be included in the objectives of the global climate. BYD already implemented the electric buses and taxis in cities of Shenzhen, Los Angeles and Santiago when it comes to the care of the transport piece of the populace. The cars of the type are linked to the central management systems of the fleet management that operates with the help of AI in order to research the efficiency of the route, the battery level, and the state of the car. This is not only reducing emissions but it also enhanced the overall mobility of the cities as the day when the traffic was stuck on transportation became a phenomenon of the past.BYD uses this ideology of a circular economy through their reuse and recycling of its battery products. Spent batteries either find use in permanently fixed-location energy storage or are reprocessed to recover valuable constituents, which are recycled to the manufacture. In general, the battery lifecycles are being monitored with the help of blockchain-based systems to establish recycling transparency. Creditably, the synergy of BYD presentation is incorporated in the vertical integration, frequency processing, and clean energy utilization. BYD is also cost-effective, responsible and environmental friendly all over the world because it has incorporated data and technology to operate all processes of the supply chain.

Case Study 3: NIO

A good example is NIO, a Chinese electric vehicle company that has become known because it focuses on innovation and user experience in the electric vehicle market. Battery-as-a-Service (BaaS) model is the core component of the strategy of NIO; it separates the purchase of batteries and a car. It allows its users to replace batteries rather than recharging them, which reduces downtimes, which extends battery life, as well as circular economy activities. BaaS is based on the assistance of such enhanced technologies as cloud computing, IoT, and AI. All NIO vehicles and batteries are equipped with IoT sensors to check whether the battery is in a good condition, its temperature, and consumption patterns. It is relayed into the centralized systems whereby the performance is analyzed and the needs of the maintenance anticipated by the AI. Also the battery at the end of service life can be recalled into circulation to be repaired and refurbished or melted in order to increase its useful life and therefore limit wastes. The NIO battery swap facilities highly automated. The process uses AI and robotics and only takes less than five minutes. Not only does it remove range anxiety when using an EV it also decentralises the charging demand and helps to take pressure off the electricity grid in and around cities. Renewable Energy is used to power these stations in which more of the carbon footprint of the services that NIO has to offer is reduced. Supply chain sustainability in the company is also data-based. NIO also follows the source of the raw materials, like the lithium and cobalt by collaborating with partners and using digital platforms to establish that the materials were ethically mined and processed. The sourcing credential is also in the trial to have blockchain on the records of the credential so as to have transparency in procurement. To achieve continuous improvement of software systems, navigation power and battery performance, the NIO automobiles receive over-the-air (OTA) updates. This does not only help in a user experience but there is less need to schedule an appointment to come and get services physically hence saving resources and energy. More broadly, NIO is designing an intelligent community around the Nio House social centres, digital venues and research into autonomous driving. The innovations are also geared to the objectives of urban sustainability such as the supply of shared mobility, intelligent transportation systems, and reduction of emissions in densely populated societies. Since incorporating the smart battery services into the Buddha-like already AI-based ecosystems and sustainable supply chain management approach to avoid greenwashing tendencies, NIO is constructing a scalable technology-based structure of sustainable mobility.

Case Study 4: May Mobility

May Mobility is a company headquartered in the United States that focuses on autonomous electric shuttles, which are meant to change urban transportation. The business focuses on restaurants and campuses with high frequency and short trips that replace the conventional way of transport with the new one that uses technology. It has cars that have AI, LiDAR and fusion sensors and this will help it drive safely without the intervention of a human driver. The green connotation of May Mobility can be duplicated in its mission of shared, electric and Autonomous solution. By exempting the private cars in the high traffic cities and introducing the electric shuttles to replace the cars, the enterprise partners in the process of reducing the greenhouse gas emissions, traffic jams, and noise to occur in the high-traffic cities. The system includes cloud-based monitoring of each shuttle in real-time mode and processing the information about vehicles and passengers as well as issuing alert messages on the basis of AI. Under these pieces of information, the routes and performance and energy efficiency of vehicles will be able to be streamlined.

At least when it comes to the supply chain, May Mobility collaborates with manufacturers, which is not part of its environmental and ethical sourcing strategy. Other components that may be chosen depending on whether they are durable and can be recycled to a circular economy can be illustrated in choice of battery and frame of the vehicle etc. Block chain is also a facility being pondered at by the company in the effort to achieve transparency in its supply chains. Autonomous tech stack is also among the ways in which May Mobility certainly stands out because it is solely a Multi-Policy Decision Making (MPDM) system. The capability of making its own secure and context-sensitive decision is based on the data of live sensors, map and traffic. It does not only make the driving significantly safer, but also allows to save energy (or electricity) as it will no longer need to burn it (or battery) when it makes extra stops or is in idle time, or even shares the wrong direction. The next level of involvement with the smart city infrastructure adds more to the May Mobility strategy. One more example of how the company cooperates with municipalities is in the introduction of the services that can align with the priorities of the urban planning, e.g., a decrease in carbon dioxide release, and the maximum utilization of connections to the public transportation system. The scalability of this model has been proved in pilot projects in Ann Arbor, Grand Rapids, and Arlington, which frequently displace low-ridership bus routes with on-demand, electrified shuttle networks. The solution by May Mobility is also all-inclusive. Its driverless shuttles are also available to those with physical limitations as well as to the transit deserts where conventional transportation is not readily available. This places sustainability in line with social equity- which is becoming a strong pillar in contemporary planning of cities. Demonstrating how a combination of autonomous EV deployment, real-time data analytics, and inclusive design can transform urban mobility into being clean, safer, and efficient, May Mobility is a positive example to the modern urban environment.

Case Study 5: MOBI

MOBI Mobility Open Blockchain Initiative (MOBI) is a nonprofit consortium that has been working on the use of blockchain, distributed ledgers, and other Web3 technologies to redefine the future of transportation and supply chains. Being a multisided consortium with its founding members, including BMW, Ford, GM, and Honda, MOBI holds a unique role in the development of interoperable digital infrastructure that makes the automotive ecosystem sustainable. The other issue that lies among the most important innovations introduced by MOBI is the problem of the Vehicle Identity (VID) Standard: it is the use of blockchain architecture adapted as a system that is used to encode unique digital identities into the cars. The technology allows an inventory of a vehicle, like manufacturing date, history of servicing, emission index and change of ownerships to be tracked in real-time safely. MOBI is enabling greater transparency, efficient and more environmentally friendly lifecycles of vehicles, by decentralization and digitization of this information. MOBI is also innovating on supply chain traceability based on the use of blockchain in authentication of the source of raw material used like rare earth metals. The consortium has already been developing the systems where all quantities of the processed materials could be tracked to the blockchain-confirmed origin in partnership with miners, manufacturers, and logistics providers. This assists in complying with the environmental laws along with the help of making responsible sourcing easy. In the case of smart mobility, MOBI is developing the concept of Mobility-as- a Service (MaaS) with blockchain incorporated that will enable seamless and shared transport mobility experiences. This involves automatic payment, free pricing as well as the carbon credit systems. As an example, the drivers using shared EV or using a public transport can get tokenized rewards, fostering sustainable travel behavior. Additionally, Trusted Trip Protocol by MOBI promotes green mobility: as it helps vehicles demonstrate that a trip has been made through secure, verifiable data. This is necessary in the programs that include congestion pricing, carbon tracking or government subsidies. It is also used to assist cities in gathering the correct data on emissions and use to guide policy and infrastructure developments. Through blockchain technology, MOBI promotes the creation of a decentralized and green network of mobility and crosses organizational boundaries in terms of data privacy, transparency, and interoperability. The given type of collaboration drives the shift toward clean, efficient, and shared transport systems. To conclude, MOBI is not a conventional motor industrial player, but itself a multi-sectoral driver of sustainable digital transformation of mobility. It enables both the public and the private to work towards decarbonizing the transport system and building ethical and resilient supply chains globally, through its standards and tools.

CONCLUSION

The use of the latest technologies, including blockchain, the Internet of Things, artificial intelligence (AI), and smart energy systems, has become one of the key factors impacting the dynamics of sustainability in the automotive sector. Such firms as BMW, BYD, NIO, May Mobility, and MOBI demonstrate that innovation may supplement the ecological responsibility ideology to develop the more transparent, efficient, and environmentally friendly organizations. These companies show that it is not only a competitive advantage but also a way to become climate resistant and sustainable mobility. The case studies indicate that, technology is not the solution but the strategic use of the tools in ethical, transparent and circular organisational settings that create maximum value. Whether by means of batteries reuse or the integration of renewable energy, through predictive maintenance powered by AI and verifying the supply chain through blockchain technologies, all the companies play their role in the wider sustainability discussion. The Agenda 2030 and the Paris agreement are some of the moves that demonstrate that sustainable transport is finally the new normal in transportation networks around the world.

Recommendations

Embrace Interoperable Digital Infrastructure: Car makers and all supply chain players ought to adopt interoperable platforms (such as the type being promoted by MOBI) that can support traceable, tamper free exchange of information and data throughout the supply chain. This is necessary to get transparency of supply chains and ESG compliance.Develop Smart Battery Lifecycle Management: As there has been an impact on the environment due to the use of EV batteries, the comapnies should adopt the developed models such as NIO, BYD, etc. that apply AI and IoT to keep track of battery health, allow second-life applications, and are responsible in recycling end-of-life batteries. Implementing Scale MaaS Solutions: Public transport agencies and privately operated transport systems ought to implement shared autonomous systems using electric vehicles, such as shuttle networks provided by May Mobility to mitigate road traffic, environmental risks, and unequal access levels to transport. Encourage integration of renewable energy: Businesses are encouraged to produce using clean energy and in production factories such as those of BYD which have solar powered factories. This has a great influence of carbon footprint reduction and improvement of brand reputation. Foster Cross-Sector Coordination: Sustainable transformation will have to be led by the cooperation of OEMs, technology startups, regulators and city planners. Consortia such as MOBI can be seen as blueprints of collaborative innovation on the industry-wide level.

Implications in the Practical and Social Implications

Practical Implications

The sustainability of automobiles due to application of superior technologies results in stronger and adaptive supply chains. Predictive analytics and real-time data enhance maintenance scheduling, waste-free material use and efficiency of production. Blockchain minimizes frauds, enables ESG reports, and ensures regulations of the transnational compliance. The resulting cost reductions, product recall minimization and improvement of the supply chain in times of disruptions that impact the world are happening due to these improvements.Moreover, AI-enabled solutions allow eliminating routine work, minimizing manipulations associated with error, and allocating resources to solve strategic tasks, including innovation and long-term planning. Among the other opportunities that will come to business as busses become more electricized is additional revenue and customer loyalty EV adoption and customer loyalty EV along with circular battery practice, and shared mobility.

Social Implications

The future technologies of sustainable mobility are quite influential socially. Firstly, the transition to the clean transportation will also have a direct impact on the air quality and human health, in case of overpopulated cities. Second, by the clever mobility solutions, the access disparities in transport of underserved groups will be bridged: e.g. rural population, elderly users or disabled users. Corporate responsibility in procurement will translate into fair working conditions and global environmental justice in the locations that have the mineral reserves and the transparent world of the digital environment makes it inclusive and accessible to the authorities. Lastly, there is also the opportunity to control and spy on consumers through the technologies, which, in turn, has a favorable ecological outcome, thus, establishing the culture of sustainability. As the user, he or she may co-authors the data-sharing eco-systems (e.g. via an BaaS or MaaS business model) and can thus make the more responsive, adaptive and green mobility solution.

REFERENCES

Aggarwal, S. (2024). Enhancing supply chain sustainability: The role of blockchain and digital technologies. *International Journal for Multidisciplinary Research, 6*(5), 28451. Advance online publication. DOI: 10.36948/ijfmr.2024.v06i05.28451

Alsaleh, T., & Farooq, B. (2024). Simulation models for sustainable, resilient, and optimized global electric vehicles supply chain. *arXiv*.

Anastasiadou, K. (2021). Sustainable mobility driven prioritization of new vehicle technologies, based on a new decision-aiding methodology. *Sustainability (Basel), 13*(9), 4760. DOI: 10.3390/su13094760

Bueno-Pascual, F.-E. (2024). *Forces transforming transport and logistics into smarter sustainable systems*. IntechOpen. DOI: 10.5772/intechopen.1005001

Cao, H., Söderlund, H., Derspeisse, M., & Johansson, B. (2023). Exploring the current applications and potential of extended reality for environmental sustainability in manufacturing. *arXiv*.

Chester, M. (2015). Can disruptive technologies, on-demand mobility, and biofuels improve transportation environmental sustainability? A review of recent research. DOI: 10.1007/s40518-015-0031-z

Deloitte. (2025). *2025 Global automotive consumer study: Key trends in mobility and sustainability*. Deloitte Insights. https://www2.deloitte.com/us/en/insights/industry/retail-distribution/global-automotive-consumer-study.html

Dias, R., Irfan, M., Galvão, R. M., & Morão, A. (2024). Analyzing persistence in sustainable indices and global stock indices: Testing the market efficiency. In *Issues of sustainability in AI and new age thematic investing* (pp. 179–201). IGI Global. DOI: 10.4018/979-8-3693-3282-5.ch011

El Jaouhari, A., Arif, J., Samadhiya, A., Kumar, A., & Garza-Reyes, J. A. (2023). An environmental-based perspective framework: Integrating IoT technology into a sustainable automotive supply chain. *Benchmarking*. Advance online publication. DOI: 10.1108/BIJ-05-2023-0322

Elassy, M., Al-Hattab, M., Takruri, M., & Badawi, S. A. (2024). *Intelligent transportation systems for sustainable smart cities*. Transport Engineer. DOI: 10.1016/j.treng.2024.100252

Esiri, A. E., Kwakye, J. M., Ekechukwu, D., Ogundipe, O. B., & Ikevuje, A. H. (2023). Assessing the environmental footprint of the electric vehicle supply chain. *Magna Scientia Advanced Research and Reviews*, *8*(2), 219–227. DOI: 10.30574/msarr.2023.8.2.0099

Fraga-Lamas, P., & Fernández-Caramés, T. M. (2024). A review on blockchain technologies for an advanced and cyber-resilient automotive industry. *arXiv*.

Grand View Research. (2024). *Smart mobility market size and share report, 2024–2033*. https://www.grandviewresearch.com/industry-analysis/smart-mobility-market-report

Irfan, M., Durrah, O., Chidambaram, N., & Singh, K. A. (2025). Green blockchain mechanisms: Enhancing transparency and efficiency in sustainable finance with DeFi solutions. In *Energy efficient deep learning approaches in IoT fog and green blockchain revolution* (pp. 211–232). IGI Global. DOI: 10.4018/979-8-3373-0300-0.ch008

Kashem, M. A., Shamsuddoha, M., & Nasir, T. (2024). Sustainable transportation solutions for intelligent mobility: A focus on renewable energy and technological advancements for electric vehicles (EVs) and flying car. *Preprints*. DOI: 10.20944/preprints202405.1677.v1

Kismawadi, E. R., Irfan, M., Haron, R., & Masitah, M. (2024). Bridging finance and sustainability: Islamic financial institutions' contributions to the transition to sustainable energy. In *Issues of sustainability in AI and new age thematic investing* (pp. 138–161). IGI Global. DOI: 10.4018/979-8-3693-3282-5.ch009

KPMG. (2024, September). The future of automotive supply chains: Connected, collaborative and resilient. *KPMG Insights*. https://kpmg.com/in/en/blogs/2024/09/the-future-of-automotive-supply-chains-connected-collaborative-and-resilient.html

KPMG International. (2024). *Supply chain trends 2024: The digital shake-up*. https://kpmg.com/xx/en/our-insights/ai-and-technology/supply-chain-trends-2024.html

Krishnan, R., Phan, P. Y., Krishnan, S. N., Agarwal, R., & Sohal, A. (2024). Industry 4.0-driven business model innovation for supply chain sustainability: An exploratory case study. *Business Strategy and the Environment*. Advance online publication. DOI: 10.1002/bse.3970

Malcalm, E., Asiedu, E., Irfan, M., & Awudu, I. (2024). Strengthening economic relationships and service supplier engagement: Pathways to indigenous inclusivity and sustainable development. In *Future of indigenous inclusivity: Unlocking finance for economic development* (pp. 129–149). IGI Global. DOI: 10.4018/979-8-3693-7924-0.ch007

Marques, N., Campos, G., & Cavalcante, M. (2024). Driving towards a sustainable future: Leveraging connected vehicle data for effective carbon emission management. *SAE Technical Paper Series*. DOI: 10.4271/2023-36-0145

Masoumi, S. M., Kazemi, N., & Abdul-Rashid, S. H. (2019). Sustainable supply chain management in the automotive industry: A process-oriented review. *Sustainability (Basel)*, *11*(14), 3945. DOI: 10.3390/su11143945

McKinsey & Company. (2025, April 22). *New twists in the electric-vehicle transition: A consumer perspective*. https://www.mckinsey.com/features/mckinsey-center-for-future-mobility/our-insights/new-twists-in-the-electric-vehicle-transition-a-consumer-perspective

Pérez-Moure, H., Lampón, J. F., Velando-Rodriguez, M.-E., & Rodríguez-Comesaña, L. (2023). Revolutionizing the road: How sustainable, autonomous, and connected vehicles are changing digital mobility business models. *European Research on Management and Business Economics*, *29*(3), 100230. DOI: 10.1016/j.iedeen.2023.100230

Riofrío, C. A., Klepsch, J., Finžgar, J. R., Kiwit, F., Hölscher, L., Erdmann, M., Müller, L., Kumar, C., & Achari Berrada, Y. (2024). Quantum computing for automotive applications. *arXiv*.

Shamsuddoha, M., Kashem, M. A., & Nasir, T. (2024). A review on Transportation 5.0: Advancing sustainable mobility through intelligent technology and renewable energy. *Preprints*. DOI: 10.20944/preprints202411.2043.v1

Shiva, C. K., Vedik, B., & Manoharan, G. (2024). *Sustainable smart transportation*. IGI Global., DOI: 10.4018/979-8-3693-2373-1.ch017

Soni, A., Shaikh, Z. H., Irfan, M., & Bijja, V. (2024). Indigenous empowerment as a catalyst for sustainable development. In *Future of indigenous inclusivity: Unlocking finance for economic development* (pp. 253–274). IGI Global. DOI: 10.4018/979-8-3693-7924-0.ch012

S&P Global. (2025, July). *2025 connected car study: Consumer priorities on connected services*. https://www.spglobal.com/automotive-insights/en/blogs/2025/07/connected-car-study-consumer-priorities-on-connected-services

Vermesan, O., John, R., Pype, P., Daalderop, G., Kriegel, K., Mitic, G., Lorentz, V., Bahr, R., Sand, H. E., Bockrath, S., & Waldhör, S. (2021). Automotive intelligence embedded in electric connected autonomous and shared vehicles technology for sustainable green mobility. *Frontiers in Future Transportation*, *2*, 688482. Advance online publication. DOI: 10.3389/ffutr.2021.688482

Xu, L., Mak, S., Proselkov, Y., & Brintrup, A. (2024). Towards autonomous supply chains: Definition, characteristics, conceptual framework, and autonomy levels. *Journal of Industrial Information Integration*, *42*, 100698. Advance online publication. DOI: 10.1016/j.jii.2024.100698

Chapter 3
Sustainability Certifications:
Tools for the Transformation of Industry

Sandeep Gupta
https://orcid.org/0000-0002-3734-3723
Graphic Era University, Dehradun, India

Devendra Kumar Tiwari
https://orcid.org/0009-0004-1408-0248
United Institute of Management FUGS, Prayagraj, India

Chandra Shekhar
https://orcid.org/0009-0003-6394-3444
United Institute of Management FUGS, Prayagraj, India

Kiran Saroj
https://orcid.org/0009-0003-7968-1899
United Institute of Management FUGS, Prayagraj, India

ABSTRACT

This chapter discusses the role of sustainability certifications in driving the adoption of sustainable practices across various industries. It outlines the significance of these certifications in promoting responsible behaviour and their global impact on industrial transformation. The chapter categorizes different certification types, including environmental and social responsibility certifications, and presents case studies from sectors such as manufacturing, retail, construction, and tech. It examines the influence of certifications on supply chains, product development, and innovation while also addressing regulatory compliance and industry incentives.

DOI: 10.4018/979-8-3373-4495-9.ch003

Furthermore, it anticipates future trends, emphasizing the role of digital tools and AI, and highlights the integration of certifications within the circular economy. The concluding perspective focuses on the sustainability landscape projected for 2030 and the vital importance of certifications in industrial sustainability.

INTRODUCTION

Given the climate crisis, environmental degradation, and social inequities that are plaguing the world, the sustainability of industrial sectors worldwide is being called into question. Increasingly, consumers, investors, government institutions, and third parties are holding businesses accountable for their contributions to the development of a sustainable future. Within this context, sustainability certifications have arisen as fundamental tools for industry change.

Sustainability certifications function as independent validations that verify an organization's adherence to established environmental, social, and governance (ESG) standards. Rather than serving as a superficial label, these certifications provide measurable and transparent proof of a company's sustainability performance (Daddi et al., 2015; Dharmarathna et al., 2025). For businesses, particularly those dependent on external service providers such as data centers, aligning corporate sustainability goals with the certifications of their partners is becoming an increasingly critical step toward maintaining credibility with customers, regulators, and investors.

Although the concept of sustainability assessment has been discussed for decades, its importance has grown substantially in recent years (Fotis et al., 2025). This growth has been primarily driven by escalating concerns about resource scarcity, widespread ecological degradation, persistent social inequality, and the increasingly severe effects of climate change (Heras-Saizarbitoria et al., 2015; Kuiper & Gemählich, 2017). As a result, industries worldwide are actively pursuing recognized certifications to demonstrate compliance with sustainable development pathways. These certifications now represent more than a mechanism for regulatory compliance; they have become instruments of transformation (Gupta & Tripathi, 2015), shaping the way organizations operate internally, how consumers evaluate products and services, and how supply chains are structured and managed.

This chapter examines how sustainability certifications can act as strategic tools that drive systemic change across sectors. The discussion highlights how certifications influence business models, decision-making, and long-term planning by embedding sustainability at the core of industrial practices. By tracing their influence, the chapter underscores the central role certifications play in addressing global sustainability challenges while simultaneously reshaping the competitive landscape of international business.

Over time, sustainability certifications have shifted from being optional symbols of good practice to becoming integral requirements for firms seeking legitimacy in global markets. Early examples such as ISO 14001 (focused on environmental management systems), LEED (promoting sustainable construction and design), and Fair Trade (ensuring ethical sourcing and labor practices) were once regarded as marketing enhancements, as shown in Figure 1 (Amundsen, 2022). Today, however, these frameworks are widely accepted as essential components of industrial transformation (Tyagi et al., 2024). By compelling organizations to consider long-term impacts, such as reducing emissions, protecting workers' health and safety, and adopting socially equitable practices, certifications encourage businesses to institutionalize sustainability through structured systems and periodic external audits.

In an era where regulatory demands, market expectations, and ecological risks are evolving rapidly, sustainability certifications play a vital role in driving innovation and adaptation. Their value extends beyond risk mitigation; they enable companies to minimize ecological harm, strengthen their reputations, meet the expectations of ethically conscious consumers, and gain entry into markets where sustainability performance is a prerequisite for entry. Consequently, these certifications are no longer peripheral but are instead central to industrial resilience and future competitiveness.

Figure 1. The Different Sustainability Certifications in the Present Market

The Urgency of Industrial Transformation

The imperative for industrial transition cannot be overemphasized. We have industrial-scale production systems based on traditional, resource-intensive models, which are incompatible with the demands of sustainability (Appolloni et al., 2022; Griffiths et al., 2018). The visible environmental impacts of industrial processes, such as water pollution and GHG emissions, illustrate this incompatibility. Meanwhile, mistreated supply chains, unregulated labor systems, and poor working conditions in numerous areas of the world deepen socio-economic disparities (Borsacchi, 2025).

Demands for change originate not only from environmental and social movements but also from investors and consumers, who increasingly require greater accountability and transparency from companies. The global market for sustainable investments exceeded \$35 trillion in 2020, as the investment industry incorporated environmental, social, and governance (ESG) factors, according to the Global Sustainable Investment Alliance. This trend has compelled the industry to reassess its approach to sustainability, and certifications have become a mechanism for verifying and communicating an organization's commitment to sustainability (Johri et al., 2025).

It is said that sustainability certifications "bring productive activities back to life" by encouraging companies to transform their practices from focusing on profit to adopting global approaches that are healthier for the environment, people, and the economy in the long term. Organizations can utilize sustainability certificates to help them establish policies and procedures that may enable businesses and the global economy to generate more revenue. Global organizations, such as the United Nations, have established the Sustainable Development Goals (SDGs) to help the world achieve these objectives.

The Purpose and Objectives of This Chapter

In this chapter, various aspects of stability certificates are explored. Their impacts on organizations within sector boundaries and their effects are also included. Through analyzing several certificates and their associated requirements, the chapter explains how certification drives industrial change, fostering innovation, efficient operations, and increased accountability. This chapter aims to demonstrate, through case studies, the various ways sustainability certifications and standards have been applied across different sectors. The benefits and disadvantages of all certifications are described here. Additionally, multiple cases are examined to illustrate how firms across various industries, including production, sales, and technology, utilize certification as a tool to achieve sustainability goals and gain a competitive edge. The future of sustainability certifications will also be examined, including emerging trends, digital tools, and the evolution of expectations among consumers, investors, and regulators. The sustainability landscape is constantly evolving, and certifications are adapted accordingly. For this reason, businesses must remain flexible to succeed in the future.

Finally, at the end of this chapter, the extensive implications of stability certificates in terms of industrial change are highlighted. This will help strengthen the notion that a shared responsibility will be in place regarding stability, to shape a more durable and equitable global economy, and to collaborate with companies, authorities, and certification bodies.

SUSTAINABILITY CERTIFICATIONS: DEFINITION AND SIGNIFICANCE

Sustainability certifications are third-party validated and provide proof that a product or organization has undergone an assessment regarding the environmental impact of its processes and whether a set of criteria defined by the certifier has been met. Some of these certifications, in addition to reducing an entity's eco-impacts and enhancing its social responsibilities, serve as external confirmations that a company or product is serious about sustainability and actively engages in sustainable initiatives (Jamoussi et al., 2022). These initiatives include reducing environmental impacts, enhancing social responsibility, and improving overall sustainable performance (Lima et al., 2021).

Figure 2. Overview of Sustainability Certifications

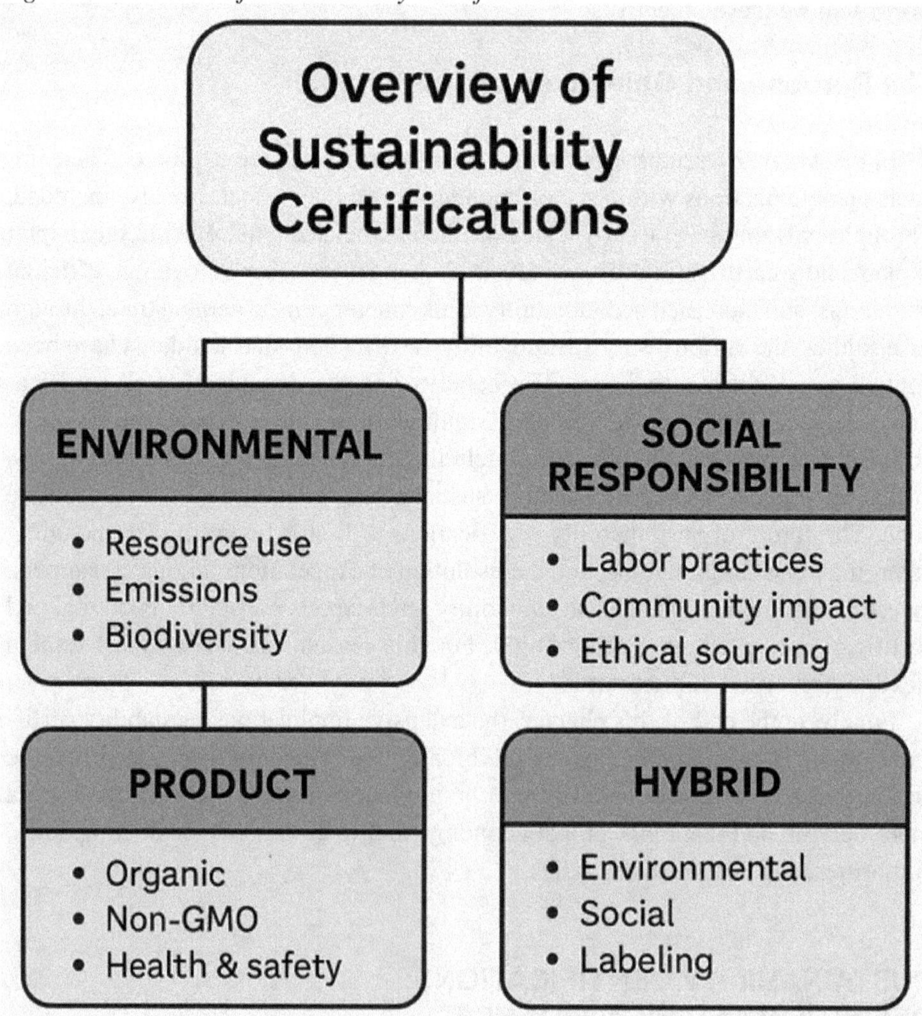

As environmental challenges and social inequities persist, sustainability certifications play a key role in driving industrial transformation. By meeting established standards, companies demonstrate a genuine commitment to sustainability that supports efficiency, innovation, and long-term success. Figure 2 illustrates the various types of sustainability certifications and their corresponding focus areas. The classes could include environmental, social responsibility, product, and hybrid certifications.

What are Sustainability Certifications?

Sustainability certification is a type of environmental certification and is also a form of Environmental Product Declaration (EPD). A step-by-step process for obtaining sustainability certification is illustrated in Figure 3, which includes a flow diagram. This type of certification process sets certain criteria, which may include (but are not limited to):

- The nature of the materials used in the production of the product (e.g., recycled or biodegradable materials)
- Energy consumption and sources (e.g., use of renewable energy)
- Water use and water conservation measures
- Carbon emission levels
- Environmental impact of the production process
- Worker working conditions and social responsibilities

The purpose of this certification is to ensure that the product or service is environmentally responsible and adheres to the principles of sustainable development (Klooster, 2006). Some of the most recognized sustainability certifications include:

ISO 14001: Global Environmental Management Standard:

- Helps organizations reduce ecological footprint.
- Promotes harmony with nature.

Fair Trade Certification: Fairness Assurance:

- Ensures fair compensation for farmers and workers.
- Promotes safe, sustainable work conditions.

LEED (Leadership in Energy and Environmental Design):

- Rewarding high-efficiency, water-conserving buildings.

B Corp Certification: Balances purpose and performance:

- Verifies social and environmental impact, transparency, and accountability.

The Significance of Sustainability Certifications in Industrial Transformation

Sustainability certifications benefit individual businesses and also help establish standards for entire industries. These certifications foster trust among businesses, consumers, and regulators, driving positive change through self-analysis of environmentally approachable practices.

Driving Change in Operations and Practices

Sustainability certifications compel companies to assess their own practices critically. However, this process is usually painful. However, these new business practices are crucial for both the environment and the company's bottom line. For example, if the organization is also seeking to achieve ISO 14001, it will need to establish and maintain an Environmental Management System (EMS) to meet this goal. These initiatives include reducing waste, optimizing resources, and enhancing recycling programs, which ultimately lead to direct operational efficiencies. Figure 4 presents a bar chart that illustrates the impact of ISO 14001 certification on key operational metrics, including energy consumption, waste reduction, and cost savings, before and after certification. A company seeking LEED certification for a building must use energy-efficient systems, renewable energy technology, environmentally friendly construction materials, and other measures that will yield a high, long-term return on investment.

Figure 3. The Certification Process

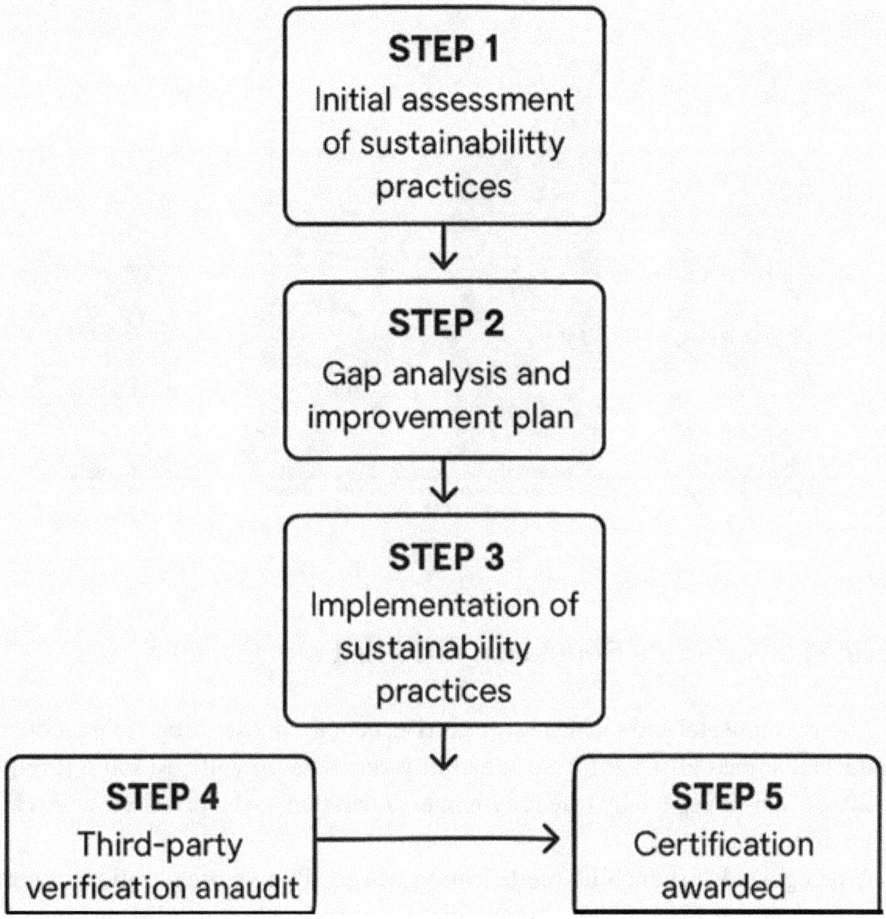

Figure 4. Impact of ISO 14001 Certification on Operational Efficiency

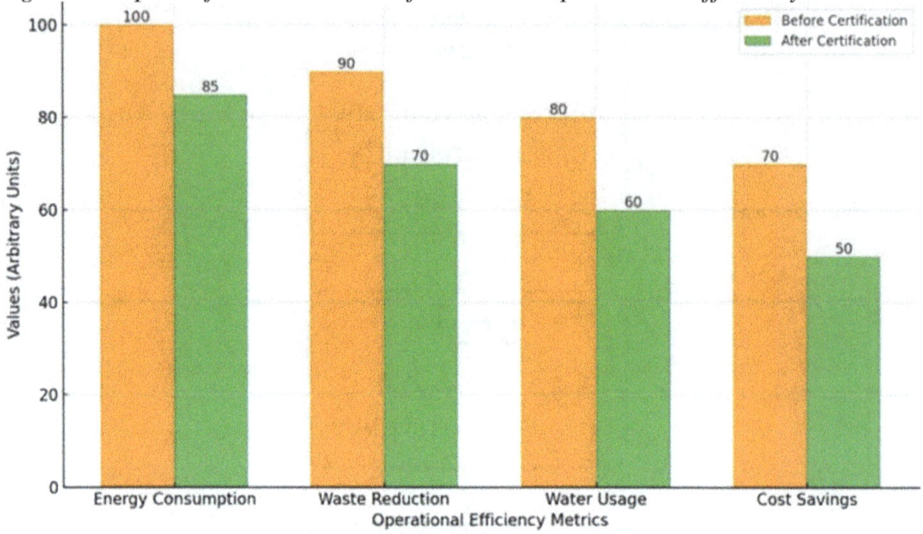

Building a Corporate Reputation and Trust with Consumers

As environmental and social awareness rise, people choose sustainable products. Products that pass HACCP for Sustainability certification will cost more (Prell et al., 2020). This is especially true for younger generations, who prioritize ethics and the environment.

A recognized sustainability certification can really give organizations a competitive advantage by confirming that their claims about sustainability are genuine. It communicates to consumers that the company is serious about its environmental and social responsibilities, which enhances the brand and cultivates consumer loyalty. For example, when a product is Fair Trade certified, it guarantees that it has been sourced responsibly, attracting ethical consumers who are keen on making responsible choices.

Accessing New Markets and Business Opportunities

As the world becomes increasingly intertwined with one another, responsible business practices are forcing organizations to realise, through pressure from investors, consumers, and even governments, that everything they do is being watched. Many new markets can be accessed only with the minimum standards of sustainability. Obtaining a sustainability certification not only expands market access opportunities but also makes regulatory access easier. For example, organizations that achieve

ISO 14001 or B Corp certifications often have an easier time getting government contracts or partnerships with other sustainable companies that ask their suppliers to meet a certain level of sustainability (Poponi et al., 2019).

In addition to this, the demand of large corporations and retailers for suppliers to have sustainability standards prior to doing business is increasing. Therefore, a large merchant like Walmart may opt not to do business with suppliers unless they have ISO 14001 accreditation or are in compliance with specific environmental standards. Such certifications will make larger supply chains accessible to SMEs, providing a critical on-ramp as they seek to scale.

Complying With Regulatory Requirements

Pipelines are subjected to monitoring from global regulatory assessments that are designed to reduce the impact of climate change, environmental degradation, and breaches of human rights (Marchi, 2022; Sharma et al., 2024). Stability certificates help companies comply with these rules and prepare for future regulatory requirements.

ISO 14001 certifications demonstrate compliance with environmental regulations, reducing penalties, lawsuits, and negative publicity. They also allow companies to anticipate future regulations, promoting sustainability. Obtaining certifications allows firms to avoid penalties, lawsuits, and unwanted press while also becoming the most sustainable company in their field.

Contributing to Global Sustainability Goals

Sustainability certifications help ensure that corporate actions align with international frameworks, such as the United Nations' Sustainable Development Goals (SDGs), thereby contributing to the global sustainability agenda (Rodríguez-García et al., 2023). Adopted in 2015, these goals provide a common framework for all countries aimed at making the world more sustainable and fairer by 2030.

Through the acquisition of various sustainable certifications, related to the Sustainable Development Goals (SDGs) outlined below, businesses can showcase the value they add to some of the most important sustainable development goals. These are a subset of the things that are part of the broader goal of:

- **Goals 12** (Responsible Consumption and Production): Numerous certifications, such as Carbon Trust and Cradle to Cradle, which are built on responsible production practices and recycling initiatives, which eventually reduce waste and pollution.

- **Goal 13** (Sustainable Development): Climate Action: Certifications like ISO 14001 contribute to the effort to mitigate climate change by providing assistance to organizations in their efforts to reduce the amount of carbon they emit (Shukla et al., 2010).
- **Goal 8** (Decent Work and Economic Growth): Social responsibility certifications like SA8000 improve working conditions, pay, and labor market ethics.

By aiding industrial sectors to make concrete steps towards attaining these global goals, these certifications help in building a better global economy.

Challenges and Limitations of Sustainability Certifications

Despite the many advantages of sustainability certifications, there are also some drawbacks and limitations to their implementation. Certifications are an important tool for sustainable business practices. From this, we can lead to long-term changes. However, this process is more time-consuming and costly. It can drive operational transformations. The first major drawback to sustainability certifications is their costs, much of which are upfront investments (Velaoras et al., 2025). Certifications require internal evaluations, reform to improve sustainability, and investments in new sustainable technologies. The initial investment fees for certifications can be very expensive for small businesses or entrepreneurs, making them difficult to afford. The other problem with certifications is complication.

There are hundreds of different certifications available across various industries for different purposes. Thus, companies must consider their options and choose between different certifications that vary significantly in terms of cost, scope, and criteria. The issue related to complexity is the lack of standardization and criteria, as some companies may seek to game the system and secure a sustainable stamp, even if their practices are only marginally better, Mitra. Despite this difficulty, businesses and consumers can be confused by the ambiguity surrounding the definition of "sustainability." In conclusion, certifications are useful tools on the path to sustainability, although they also present significant obstacles for businesses.

TYPES OF SUSTAINABILITY CERTIFICATIONS

Sustainability certifications are essential instruments that contribute to the transition of industrial sectors through the adoption and showcasing of sound practices within the environmental, social, and economic dimensions. These certifications support firms in formally implementing their sustainability practices and reflect their commitment to sustainable practices. There are many different types of certifications.

But this is very difficult to choose the correct certification for one's own industry. However, the goal of each path is still the same: to make the world a better place.

In this section of the book, we provide a brief overview of sustainability certifications. The primary focus will be on the most common certifications related to environmental impact, social responsibility, product sustainability, and similar areas. It will incorporate significant case studies that will demonstrate how worship businesses utilize these credentials to advance their efforts and increase their accuracy.

Environmental Certifications

Environmental certifications are primarily focused on minimizing the negative environmental impact of a company, product, or service. Such certifications typically relate to energy efficiency, waste management, water conservation, carbon emissions, and the sustainable management of resources.

ISO 14001: Systems for Managing Environmental Matters

A recognized sustainability certification can really give organizations a competitive advantage by confirming that their claims about sustainability are genuine. It communicates to consumers that the company is serious about its environmental and social responsibilities, which enhances the brand and cultivates consumer loyalty. For example, when a product is Fair Trade certified, it guarantees that it has been sourced responsibly, attracting ethical consumers who are keen on making responsible choices.

Key Areas of Focus:

- Waste reduction
- Energy effective
- Reduce wastages
- Control of pollution
- Managing environmental risk

Energy Star Certification

Energy Star is a government-backed program helping businesses and individuals protect the environment through superior energy efficiency. Its buildings, industrial facilities, and products that earn the Energy Star label meet strict energy efficiency specifications set by the U.S. Environmental Protection Agency (EPA) and the Department of Energy (DOE). The certification helps organizations reduce energy use, lower operational expenses, and support more ambitious environmental objectives, including reductions in greenhouse gas emissions.

Key Areas of Focus:

- Efficient buildings and energy appliances
- Lowering greenhouse gas emissions
- Energy efficiency cost savings
- Environmental regulation compliance

Case Study:

ISO 14001 certification was attained for a group of manufacturing plants in a global electronics company. By implementing an EMS, energy consumption was cut by 18% and water consumption by 25%, resulting in considerable savings through lower bills and a more sustainable production system. This helped the company further establish itself as an environmentally friendly frontrunner in the sector.

Social Responsibility Certifications

Social responsibility certifications, including certifications addressing the well-being of employees, fair labor practices, and ethical behavior within a global supply chain. These certifications typically concern payment of a living wage, health and safety of workers, non-discrimination policies, and respect for human rights.

SA8000: Social Accountability

SA8000 certification is one of the most widely recognized and respected social accountability standards. This is to make sure companies respect human rights and fair labor conditions, to protect workers from abuse, and to promote ethical conduct up and down the supply chain.

Key Areas of Focus:

- Fair wages
- Safe working conditions
- No child or forced labor
- Right of free assembly
- Non-discriminatory

Fair Trade Certification

Fair Trade is a comprehensive approach to trade that seeks to achieve greater equity in international commerce. The certification centers on fair trade, sustainable development, fair wages, and environmental consciousness. Figure 5 shows a pie chart, which shows the benefits of Fair-Trade certification in terms of increased wages, improved working conditions, better community development, and environmental sustainability.

Key Areas of Focus:

- Fair pay for farmers and workers
- Producer–Organization Direct Trade
- Committed to environmental sustainability
- Aiding those less fortunate

Case Study:

A popular coffee company also embraced the Fair Trade mark and partnered with coffee growers in Latin America. The certificate also guaranteed fair pay, better working conditions, and access to education for workers. It created long-term relationships between Rijk Zwaan and growers, making the supply chain sustainable for all involved.

Figure 5. Comparative Benefits of Fair-Trade Certification

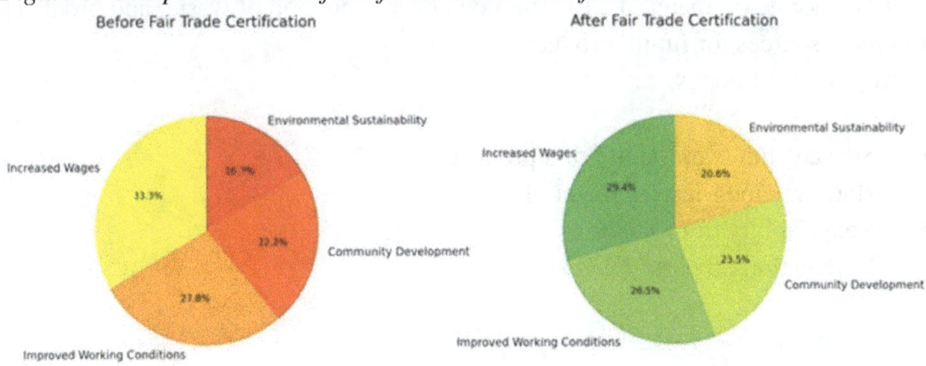

Product Certifications

Product certifications are issued for individual products or services, showing that a specific product or service fulfills specified sustainability criteria. Such certifications are often concerned with assessing the lifecycle impact of a product in terms of the environmental and social implications of the product from design and sourcing all the way through manufacturing, to use, and end-of-life disposal.

LEED: Leadership in Energy and Environmental Design

LEED is an internationally recognised certification for high-performance in sustainability buildings and different infrastructure initiatives. The certification assesses the environmental impact of site development, including water, energy, and material consumption, and encourages sustainable practices.

Key Areas of Focus:

- Eco-friendly design and construction
- Conservation of water
- Quality of the indoor environment
- Sustainable site development
- Sustainable carbon emissions

Cradle to Cradle Certification

Cradle to Cradle is based on the life-cycle of a product and encourages designers to create products that are fully recyclable or biodegradable. It promotes the use of safer and non-toxic materials, focusing on closing the loop of waste, and products become resources for future products.

Key Areas of Focus:

- Stewardship over the life of a product
- Harmless long-lasting materials
- Recyclability of products
- Elimination of waste

Case Study:

Cradle to Cradle Certified™ for a leading brand of furniture. The brand, by using materials sourced sustainably and considering recyclability, was able to develop a completely circular product cycle, which not only attracted many more environmentally conscious consumers but also helped lower material costs by reusing elements.

Other Notable Sustainability Certifications

Aside from the environmental, social responsibility, and product-focused certifications, several others span various areas of sustainability. These hybrid or sector-specific certifications are helping organizations address specific sustainability issues while promoting long-term change.

B Corp Certification

B Corp is a certification for organizations that meet high standards of social and environmental performance, accountability, and transparency. B Corps balance profit and purpose by considering the impact of their decisions on employees, customers, suppliers, the community, and the environment.

Key Areas of Focus:

- Social and environmental performance
- Transparency and accountability
- Employee benefits and fair wages
- Community engagement

Carbon Trust Standard

The Carbon Trust Standard certifies organizations that effectively measure, manage, and reduce their carbon emissions. It supports organizations in reducing their carbon footprint, enhancing energy efficiency, and contributing to global efforts to mitigate climate change.

Key Areas of Focus:

- Carbon footprint measurement
- Reduction in carbon emissions
- Energy efficiency improvements
- Environmental responsibility

Forest Stewardship Council (FSC) Certification

The FSC certification applies to forest products, ensuring that these materials are sustainably sourced. FSC-certified products are made from wood, paper, and other forest materials that meet the highest environmental and social standards.

Key Areas of Focus:

- Sustainable forestry practices
- Protection of biodiversity
- Community involvement in forest management
- Ethical sourcing of forest products

Case Study:

A large paper manufacturer achieved FSC certification for its range of paper products. By sourcing paper from sustainably managed forests, the company helped preserve ecosystems, supported indigenous communities, and gained consumer trust by offering a sustainable alternative to traditional paper products.

Emerging Sustainability Certifications

The field of certifications is always changing as per customer demand. There has been a huge increase in the number of new certifications based on the changing needs and challenges in the sustainability space. Nevertheless, not all of these have achieved a substantial market presence; some have already established a niche, while others are still in the process of doing so. It attempts to remain true to the primary themes of climate change, biodiversity loss, and the circular economy. These developments also respond to the most recent technological advancements (general-purpose technology) and the emerging environmental and social issues in this context.

Circular Economy Certifications

Circular economy: The certifications focus on business models for resource efficiency, as well as waste and product lifecycle management. The certifications incentivize companies to create products that are reusable, repairable, or recyclable, in turn, supporting the manufacture-to-factory feed loop.

Net Zero Carbon Certification

As the crisis of climate change becomes increasingly serious, zero net carbon certifications become more necessary. These are awarded to organisations that attempt to minimise their greenhouse gas emissions by offsetting them with organisations that aim to reach zero emissions directly or through the purchase of offsets.

While we currently offer the widest and strongest selection of sustainability certifications, each addresses only part of environmental, social, and economic sustainability. From environmental management systems, such as the ISO 14001, to product-based certifications like LEED and Fair Trade, certifications provide companies with a means of achieving overall sustainability while complying with international ethical, environmental, and social values and standards (Lemos & Agrawal, 2006).

By ensuring sustainability certifications are enforced, companies can not only adhere to environmental legislation but also actively add value to society by tackling social challenges, such as climate change, workers' rights, and responsible procurement. Whether they're lowering their carbon footprint, enhancing their workers' rights policies, or developing more circular products, businesses are utilizing their sustainability certifications to drive change, change that will create a more sustainable future across multiple industries and communities.

This collection of case studies presents empirical instances of organizations from diverse sectors utilizing sustainability certifications to effect change and affirm their dedication to sustainability goals.

THE ROLE OF CERTIFICATIONS IN DRIVING INDUSTRIAL CHANGE

Sustainability labels as a tool for industrial transformation. They also provide practical advice, established procedures, and credit to companies that have aligned their businesses with environmental, social, and governance (ESG) criteria. Certification is also an investment and a commitment, but it enables organizations to make a positive impact on their business practices and sustainability goals.

In this section, we will discuss how sustainability certification transforms industries, from better supply chain management to product innovation. It will also cover the pitfalls and rewards of certification, showing how companies can leverage the credibility, accountability, and competitive advantage of certification.

Certifications as Catalysts for Change

Sustainability certifications prompt organizations to reassess their operations, reevaluate strategies, and refine their business models. They typically require changes that impact how products are manufactured, how resources are utilized, and how waste is managed.

Driving Efficiency and Operational Improvements

However, to achieve this level of sustainability certification, numerous organizations must be strengthened and challenged to meet these standards. These organizations must be capable of identifying inefficiencies, shortcomings, and waste in a business from all perspectives in order to verify its sustainability. The majority of businesses will use processes that consume less energy and produce less waste to obtain certificates such as ISO 14001, as seen at Pest Control Company Millerville. Currently, a company seeking ISO 14001 certification must establish an environmental management system (EMS) that ensures all processes throughout its operations are designed to incorporate environmental considerations that conserve resources and minimize pollution. That in turn makes for a more satisfactory arrangement of work, cheaper in many cases, and most effectively minimizes wastage throughout the product life cycle, energy certifications. As soon as the e.g. is obtained, even though on average a system builder Samsung monastery system is 30% higher in air volume than using low-tech cheaperointments high volume convection cooling plus 240mm base plate * 60cm fins Full copper dual 2PLC (Productive Limit Control) power on electric start fail safe readout lead makes the PC2762 card one of our most useful ever. Obsolete now after Energy certifications (Energy Star), on the other hand, force organizations to adopt energy-efficient technologies. By doing so, they can help reduce energy consumption and save the company some costs.

Case Study

An automobile manufacturer sought ISO 14001 certification to enhance its sustainability. They identified several inefficient production lines during the certification process, where waste and energy consumption were high. In two years, the company reduced its energy consumption by 20% through improvements made to its Environmental Management System (EMS). This significantly reduced its environmental impact and also its costs. In this way, the company became more sustainable and gained a competitive advantage in a market where eco-friendly products are gaining ground.

Shaping Product Development and Innovation

Certifications aimed at sustainability impact how organizations design and shape their products, considering the entire product life cycle. The introduction of standards and certification creates incentives to ensure the use of sustainable raw materials, efficient technology, and designs that are easily recyclable. This helps in driving firms towards innovation in line with sustainability.

Innovation in Product Design

Green certifications for buildings, products, and services provide firms with the necessary benchmark to adopt a sustainability agenda. For example, LEED or Cradle to Cradle evaluates product designs that minimize waste and utilize non-toxic materials, allowing for recycling or reuse. Companies are encouraged to innovate by creating environmentally friendly and economically viable alternative materials or production methods as a means of obtaining these certifications.

Cradle to Cradle aims for a "closed-loop" lifecycle. A product that gets disposed of isn't the solution for their designers. Products need to be reusable or recyclable and certainly not destined for a landfill. This means organizations must change the entire approach of their production and materials sourcing to make their products more sustainable and circular.

Case Study

The company sought Cradle to Cradle certification for its new line of kitchen appliances. The company exclusively uses recyclable materials for all components of its products. The company has redesigned it to facilitate easy disassembly and reuse, thereby increasing its life cycle. This modification facilitated the attainment of a more sustainable product and allowed for the inclusion of a new consumer segment within the circular economy. Due to this, the company is able to decrease its dependence on raw materials. Therefore, finally, there is a reduction in production costs and a more environmentally sustainable process.

Enhancing Supply Chain Sustainability

Certifications can have a significant impact on the entire supply chain when a company adopts a sustainability certification. Then, it creates expectations for its suppliers to meet specific standards of environmental and social responsibility. In simpler terms, this means that suppliers must also be certified to the same level. This ripple effect encourages the spread of sustainable practices across different sectors, driving change that reaches far beyond the company's own boundaries.

Collaborative Supply Chain Improvement

In fact, for many certifications, especially (but not limited to) social responsibility (SA8000, etc.) or environmental performance (ISO 14001, etc.), organizations must partner with their suppliers to ensure they are performing similarly. It also promotes increased transparency in the supply chain and collaboration between organizations and their suppliers to achieve shared sustainability goals.

Sustainability certifications help companies compete less based on labor and environmental standards in a "race to the bottom" mentality and instead select suppliers based on responsible practices. Such a transformation is crucial for building more resilient and sustainable supply chains in industries such as apparel, agriculture, and electronics.

Case Study

A global retailer investigated Fair Trade certification for one of their clothing lines; as part of this process, the organization had to engage with its supply chain to prove that every supplier was offering fair labor practices. During the certification process, the company collaborated with local farmers and manufacturers in developing countries to ensure fair wages for workers, safe working conditions, and the provision of healthcare. This not only strengthened relationships within the supply chain but also enabled the retailer to enhance its reputation with socially conscious consumers.

Improving Corporate Governance and Accountability

Sustainability Certification programs help provide a framework. Many of these certifications are associated with a rigorous accountability framework that creates mandatory policies and practices within firms to support globally recognized sustainability objectives.

Accountability and Transparency

Many certifications, such as B Corp, require organizations to disclose their social and environmental impact publicly, which creates the potential for greater accountability. When a company shares data and information with relevant stakeholders, it essentially opens its doors to its customers. Moreover, that level of transparency can earn the company a great reputation.

One great example of a certification is B Corp. It's a thorough and demanding evaluation that examines how well an organization manages its governance, workplace, environmental impact, and community involvement. Companies with this certification are held accountable not just by regulators. Consumers, employees, and investors expect these companies to focus on long-term sustainability, not short-term profitability.

Case Study

Global is presumably a technology company, and after obtaining the B Corp certification, it seeks to emphasize that it is also a responsible entity. The organization recognized the flaws in its governance methodologies and opened up to public examination, thus ensuring that its products and services ran at high ethical standards. As a result, a new governance policy was introduced. This policy fosters a better relationship between corporations and their stakeholders. Moreover, it enhances consumer trust.

Competitive Advantage and Market Differentiation

A sustainability certification can easily be the differentiator between your business and your competitor in a world where consumers are rapidly preferring sustainable brands. Certifications help businesses differentiate themselves from their competitors by attracting eco-conscious consumers, investors, and business partners who value sustainability and environmental responsibility.

Brand Loyalty and Consumer Trust

As consumer concern around sustainability increases, firms with sustainable certifications may gain more brand loyalty. Certifications serve as a mark of credibility, demonstrating to the consumer that the company operates responsibly. With certifications, such as Fair Trade or ISO 14001, they help build trust and create long-term relationships, especially with consumers who want to buy from brands that share their values.

Case Study

Kering's luxury fashion label got its clothes Fair Trade certified, making the brand's ethical labour practices official. This certification enabled the brand to appeal to an increasing number of ethical consumers who are ready to pay a premium for high social and environmental standards. The company experienced an increase in sales due to increased customer loyalty and media coverage that strengthened its market position.

Sustainability certifications are effective levers to drive change in the industry. They act as sparks for operational improvements, product innovation, supply chain transparency, corporate governance, and market differentiation. Compliance with sustainability certifications enables organizations to improve their efficiency, minimize ecological and social impacts, and enhance their long-term viability. Additionally, these certifications offer organizations a clear framework for managing their sustainability activities, which not only enhances accountability and transparency but also fosters customer confidence and loyalty.

As different industries change, adapt, and innovate to meet the growing demand for sustainability, certifications will continue to play a crucial role in shaping the direction of global practices, enabling businesses to navigate the challenges of the 21st century.

CASE STUDIES: INDUSTRIAL TRANSFORMATION THROUGH CERTIFICATIONS

This is not to suggest that sustainability certifications are the sole reason for an entire industrial transformation; real-world examples must be studied in order to gain a true understanding of how sustainability certifications have a leverage effect. In this section, we present several case studies across multiple industries that demonstrate how companies have utilized sustainability certifications to optimize operations, develop more effective products, streamline supply chains, and drive innovation.

These case studies will also help us investigate the practical advantages and challenges of sustainability certifications, as well as how organizations have leveraged these instruments to align their actions with global sustainability goals. Showcasing a variety of certifications in environmental and social responsibility, the case studies will provide an overview of the rise of these tools for industrial transformation.

Case Study 1. Certification of ISO 14001 in Manufacturing

About The Company: Less than $1 Billion company || Global presence in over 20 countries || Leading global manufacturer of electronics. It designs, produces, and sells consumer electronics, home appliances, and telecommunications equipment.

Certification description: ISO 14001 is an international standard for Environmental Management System (EMS). It provides a structured approach for organizations to enhance their environmental performance by making more efficient use of resources and reducing waste.

Challenge: facing growing pressure from regulators and consumers to lower its environmental footprint, increasing energy costs, and waste disposal problems were also hurting profits.

Solution: ISO (International Organization of Standardization) 14001 certification was adopted for each production site by the model firm. This included implementing an EMS that would help them reduce energy use, cut down on waste, and achieve compliance with environmental regulations. The process involved:

- Doing environmental audits to find areas of improvement.
- Goals for quantifiable reductions in carbon emissions, energy consumption, and waste
- Empowering employees at all levels to adopt and measure sustainable policies.

Results:

Energy Reduction ¨C The company introduced technologies aimed at improving energy efficiency (LEDs, thermostats, and renewable sources, such as solar panels), which helped it reduce 15% of the energy used in its manufacturing facilities.

- **Waste Management:** Through optimized waste management processes, the company was able to lower landfilled waste by 30% and increase recycling by 50%.
- **Cost Savings:** The environmental initiatives resulted in large cost savings, especially in the form of energy bills and waste disposal fee reduction.
- **Improved image:** Certification helped the company to establish a positive reputation as an environmentally friendly manufacturer, gaining the favor of investors and customers who place more weight on sustainability.

Takeaway: No longer just ISO 14001 helped the company comply with regulations; it has also helped achieve operational efficiencies, cut costs, and improve its sustainability profile —a competitive advantage in the global market.

Case Study 2. Fair Trade Certification in Agriculture

Company: International company engaging in the retail of organic food items, including fruits & vegetables, packaged products, and drinks. It imports all products from around the world, including developing areas within Latin America, Africa, and Southeast Asia.

Certification Summary: Fair Trade helps ensure that products are sourced from suppliers that adhere to strict social, environmental, and economic standards, including fair wages, safe working conditions, and sustainable farming practices.

Challenge: The company faced increasing pressure from consumers to source ingredients ethically and ensure animal welfare. Nonetheless, many of the company's suppliers, especially in the developing world, suffered from labor abuse, substandard working conditions, and environmental damage.

Solution: After observing consumer demand for ethically sourced products, the firm pursued Fair Trade certification for its agricultural supply chain. The certification process involved:

- Evaluating every supplier on whether they were aligned with Fair Trade labor and environmental standards.
- Giving fair wages, healthcare, and safe working conditions to everyone in the supply chain — farmers and workers.
- Encouraging sustainable agriculture, such as organic farming, water saving, and waste minimization.

Results:

- **Social Impact:** The Fair Trade certification obtained by the company has helped thousands of workers in developing countries with better wages and working conditions. Premium payments earned by farmers also allowed them to reinvest in their communities.
- **Environmental Impact:** The company helped reduce waste, improve soil health with sustainable farming practices, reduce the need for pesticides and chemicals, and conserve water.
- **Boost Consumer Loyalty:** The certification appealed to ethical consumers who would pay higher prices for Fair Trade certified products. This, in turn, increased sales of the company's ethical food product lines.
- **Market Growth:** By targeting consumers who value sustainable and ethical labor practices, the company grew its market share, allowing it to stand out in an oversaturated market.

Key Takeaway: The Fair-Trade certification addresses both social and environmental issues within the supply chain, while also creating a competitive advantage through the support of an expanding group of conscious consumers. The market growth of sustainable products is shown in Figure 6.

Figure 6. Market Growth of Sustainable Products

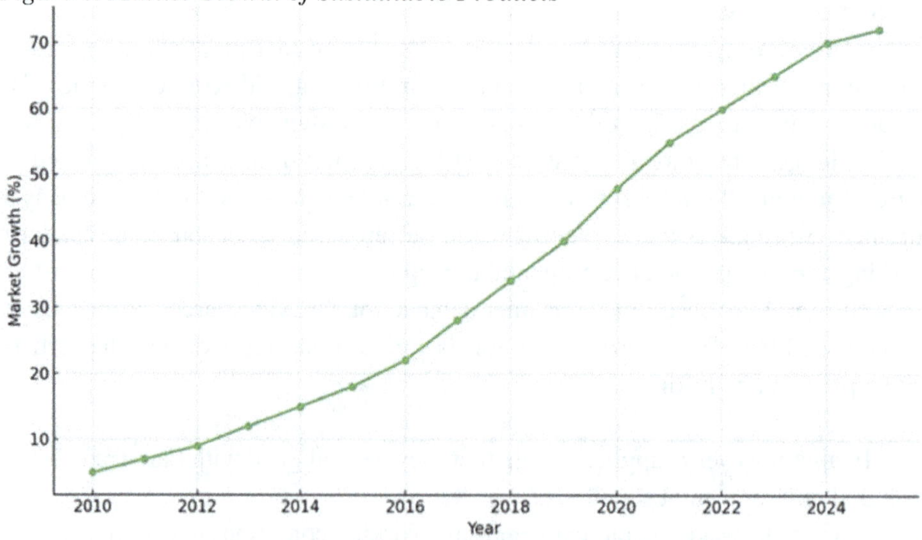

Figure 7. Global Distribution of LEED Certified Buildings

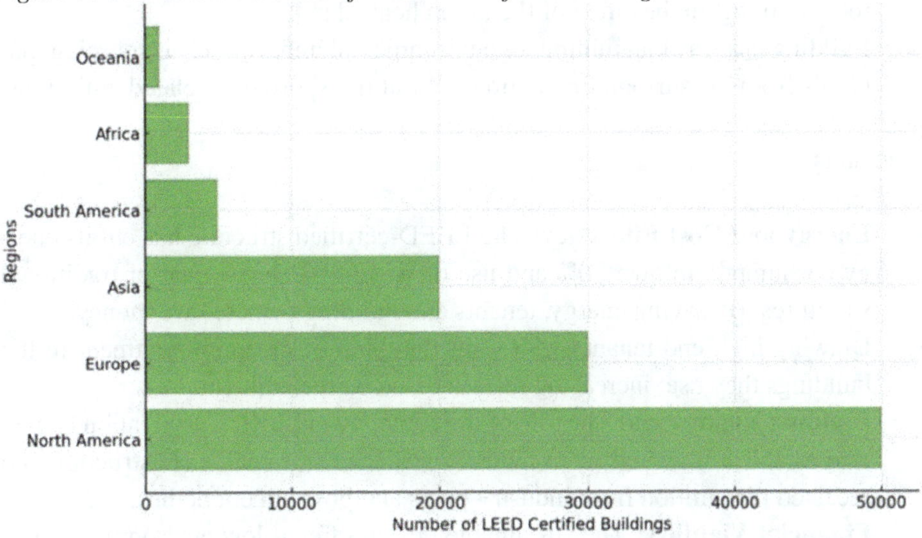

Case Study 3. LEED Certification in Construction

Company Profile: One of the biggest real estate development companies working on various large architectural investments, commercial, and residential buildings. The firm is credited for innovation in the field of sustainable urban design. Figure 7 shows the global reach and adoption of sustainable building practices.

Quick Summary of the Certification: LEED is a universal green building certification that recognizes high achievements in energy efficiency, water conservation, and sustainable site development.

Challenge: The company aimed to be an industry leader in the competitive real estate market, helping it carve out a niche with environmentally sustainable buildings. However, it was under growing pressure from the government and consumers alike to change its methods to minimize energy use and become more sustainable.

Solution: The company chose to obtain LEED certification for its new high-rise office building project. The certification process involved:

- Implementing energy conservation measures such as new high-efficiency HVACs, LED lights, or even high-performance glazing.
- Incorporating water-efficient elements in the building, including low-flow fixtures and rainwater harvesting systems.

- Sustainable site initiatives, incorporating recycled materials, designing green roofs to mitigate the effect of the urban heat island.
- Building in a way that minimizes environmental impact, such as sourcing materials from the surrounding region to limit transportation-related emissions.

Results:

- **Energy and Cost Efficiency:** The LEED-certified structure has cut its energy consumption down 30% and use of water 40% versus that of traditional structures. By saving energy, tenants and building owners save money.
- Drawing high-end tenants who want this kind of green commitment in the buildings they use; increasing market value on the bulk-end.
- **Building Reputation:** The role of the company in LEED Certification helped it to establish its brand as an industry leader in sustainable construction and received recognition from industry peers and government bodies.
- **Financial Viability:** The sustainable design offered low environmental impact and long-term cost savings for the company and tenants, maintaining financial viability.

Takeaway: LEED certification demonstrated that sustainable building practices can be effectively integrated into commercial real estate development, creating opportunities for long-term cost savings, enhanced tenant satisfaction, and increased building brand value.

Case Study 4. B Corp Certification in Technology

Employer: Well-funded tech startup creating software for small companies. It has established a reputation for product innovation and ethical practices.

About the Certification: B Corp certification is for businesses that meet high standards of social and environmental performance, accountability, and transparency.

Problem: As the business expanded, the company faced increasing challenges to ensure that its rapid growth was consistent with its social and ethical values. It aimed to convey its dedication to sustainability and establish trust with consumers and investors alike.

Solution: To formalize their commitment to sustainability, the company sought B Corp certification. The certification process involved:

- Scrutinizing and refining the internal practices of the company when it comes to the areas of duty towards employees, Diversity, and Community engagement.

- And adopted measures to lower the environmental impact of our business, switching to green energy and minimizing paper waste.
- They implemented a transparent governance structure that involved regular reporting on social and environmental performance.

Results:

- **Social Environmental Performance:** The B Corp certification boosted the company's image as a responsible tech firm and helped attract new customers who like transparent, sustainable companies.
- **Investor Confidence:** Investors were coming to B Corp certification as it provided them with assurance that the company was focused on long-term sustainability and not short-term profit.
- **Employee Retention:** Ethical practices improved company morale and employee well-being while helping with employee retention and productivity.

B Corp certification helped the company cement its ethical foundations while growing, aligning profits with a greater purpose in society and the environment.

These case studies demonstrate how sustainability certifications can be leveraged to transform industries. From environmental management systems, such as ISO 14001, to ethical labor standards with Fair Trade, sustainable building designs with LEED, and full-spectrum social responsibility with B Corp, certifications can be an effective means for organizations to secure significant and lasting improvements to their operations, products, and supply chains.

In each case, these organizations were able to compete more effectively, reduce costs, and enhance their sustainability credentials while also doing good. These changes demonstrate the ability of sustainability certifications to not only catalyze innovation and reduce risk but also to foster stronger ties with consumers, investors, and other stakeholders, leading to increased loyalty and trust. Valuably, through this understanding of the challenges and successes of these organisms, firms can have a much clearer picture of the benefits and scope of sustainability certifications and their role in industrial transformation.

FUTURE OF SUSTAINABILITY CERTIFICATIONS

Evolving industrial needs, consumer demands, and global government initiatives will shape the future of sustainability certifications. In other words, sustainability certifications will remain crucial to driving this change as the world transitions toward greater sustainability in its economy. This section will discuss some of the novel challenges and innovations shaping the future of sustainability certifications, as well as the evolving roles of these certifications within industrial transformation.

Evolving Trends in Sustainability Certifications

With industrial sectors already on the path to realizing sustainability, the need for vague, superficial, and weak certificates will end; instead, certificates should be detailed, clear, and stringent. These unfolding trends represent the transition to a more sustainable, circular, and socially responsible global economy.

Integration with the Circular economy

Different sectors and industries are realizing the potential of the circular economy. This industrial economy maximizes the value of products, materials, and resources by keeping them in use for as long as possible. Sustainability certifications will need to adapt to support the circular economy trend, as more resources will be created and consumed to meet the world's needs while preserving its resources. The initiative also advises businesses to create products that can be reused, repaired, or recycled rather than being discarded after use.

Initiatives such as Cradle to Cradle and Circular Economy 100 are already spearheading this change by instituting ways of thinking about product design and manufacturing processes that encourage recyclability, up-cycling, and closed-loop systems. The adoption of circular economy principles is expected to increase, creating opportunities for certifications to facilitate their implementation.

Carbon Neutral and Net-Zero Certifications

With the clock ticking down on climate change action, there is a growing emphasis from many organizations on achieving net-zero carbon emissions, either by reducing emissions within their own operations or by investing in carbon offset projects (Gupta & Tripathi, 2014). We will see this trend play out in future sustainability certifications, with new ones emerging or existing ones evolving to help firms measure, manage, and mitigate their carbon footprints.

As carbon reduction becomes an increasing priority, certifications such as the Carbon Trust Standard and B Corp are now requiring more stringent carbon reduction criteria. In the near future, we may also see more niche certifications that focus specifically on carbon neutrality or net-zero emissions. These certifications will help companies not only measure their scope of carbon emissions but also provide a roadmap for reducing carbon in the long term.

Digitalization and Automation of Certificate Processes

With the digitization of the entire world, sustainability certifications will integrate with digital tools and technologies. Digitalization could also possibly speed up certification, tracking, and reporting, while allowing for greater transparency in the supply chains and operations. One example includes the exploration of blockchain technology and the traceability of certified products, allowing real-time verification of sustainability claims by consumers and firms.

It would also facilitate improved sustainability metric tracking and reporting, allowing firms to measure their performance and third-party auditors to assess performance more effectively. Furthermore, the use of artificial intelligence (AI) and machine learning will enable more accurate predictions of sustainability results, thereby enhancing the accuracy of certification processes (Gupta, 2025; Sinha et al., 2023).

Blockchain technology has been utilized by a global apparel brand to track the entire supply chain of certified organic cotton products, from farm to store (Sinha et al., 2024). This ensures authenticity and transparency, allowing customers to digitally verify the sustainability credentials of their purchases.

Overcoming Challenges in the Certification Process

Sustainability certifications have been important instruments for industrial change; however, there are still significant areas that need improvement in the future so that these certifications can play a more significant role.

The Confusion and Fragmentation of Certification Standards

However, the sustainability certification landscape is complex and fragmented, making it one of the most significant challenges faced today. There are hundreds of certifications out there, each one unique in its requirements and criteria. This can be daunting for firms, particularly small and medium-sized enterprises (SMEs), which lack the resources to undergo certification.

In response to this challenge, there is an increasing demand for greater harmonization and standardization of sustainability certifications that can facilitate this process. There are ongoing efforts to develop consistent frameworks that can compete with existing comprehensive frameworks and align with global sustainability goals, such as the United Nations Sustainable Development Goals (SDGs), thereby enabling firms to provide clear and actionable direction for their activities.

One illustration of this trend is the increasing desire to couple ISO 14001 with the Global Reporting Initiative (GRI) standards and B Corp certification to better facilitate firms in demonstrating their ESG performance in relation to both environmental and social sustainability principles (González-Benito et al., 2020). With a more standardized process, certifying bodies will simplify the process, reduce confusion, and ultimately lower the costs for organizations applying for certification.

Addressing Greenwashing Concerns

While more organizations are jumping on the sustainability bandwagon, many of these same organizations misrepresent or even outright lie about the sustainability of their products or operations, a behavior scientists call greenwashing. There are instances, however, where organizations do not have a third-party certification process to support their sustainability claims, which detracts from the integrity of the certification.

To combat greenwashing, however, certification bodies will need to tighten their auditing processes and ensure greater transparency and accountability. This may involve more stringent assessments, frequent audits, and explicit documentation of the process for certification. Furthermore, companies seeking certification will need to substantiate that their sustainability practices are genuine and not merely a marketing tactic.

Low Value, High Cost, and Resource Intensive for SMEs

For a certain population of small and medium-sized enterprises (SMEs), certification may be cost-prohibitive, e.g., LEED or Fair Trade, due to the high initial capital outlay required for infrastructure and process changes. As a result, organisations that could benefit the most from such sustainability certifications may be obstructed due to this high barrier to entry.

One potential solution is to create specific certifications for SMEs with fewer requirements, lower costs, and easier implementation. These certifications would be more about incremental change than a widespread overhaul, allowing smaller firms to gradually signal their sustainability credentials upon winning certification.

Governments and financial institutions can also provide incentives, subsidies, or low-interest loans to help SMEs obtain sustainability certifications. This would create a level playing field, allowing other firms to join the sustainability wave.

Emerging Certification Trends: Impact on Different Industries

With sustainability now near the top of the agenda for governments, businesses, and consumers alike, several industrial sectors are likely to experience increased demand for specific sustainability certifications. These trends are driven by specific sustainability challenges faced by each sector, as well as rising consumer and regulatory pressures to meet environmental and social performance standards.

The Manufacturing Sector

The manufacturing sector will see an adoption of certifications that target energy-efficient practices, waste reduction, and sustainability-based production methods. Standards such as ISO 14001, Energy Star, and Cradle to Cradle will play a crucial role in facilitating the adoption of broader sustainability in manufacturing supply chains. Moreover, the increased focus on circular economy practices will drive more companies to pursue certifications that emphasize sustainability across the product life cycle, waste management, and resource efficiency.

Agriculture and Food

There is an extraordinary need for and excessive use of resources, environmental degradation, and double standards in labor practices within the agriculture and food industry. Certifications such as Fair Trade, Organic, and Rainforest Alliance will continue to play a crucial role in promoting sustainable farming practices and ensuring fair labor conditions. With the growing demand among consumers for ethically sourced food, the agricultural sector can expect an increase in animal welfare, reductions in carbon emissions, and a greater emphasis on pursuing sustainable sourcing certifications.

The Construction and Real Estate Sector

Sustainability certifications, such as LEED (Leadership in Energy and Environmental Design) and BREEAM (Building Research Establishment Environmental Assessment Method), will be further adopted by the construction and real estate industries. Such certifications shall also include green building design, energy-efficient construction techniques, and those related to low-carbon infrastructure projects. Moreover, with the upcoming trend of Smart cities and sustainable urban development, the certification market will also focus on renewable energy, smart grid, and sustainable mobility forward and backward linkage solutions.

The Technology and Electronics Sector

Certifications related to e-waste management and product lifecycle sustainability are expected to gain momentum in the tech sector, as will energy-efficiency certifications. The Energy Star certification will continue to be an important eco-label for the environmental performance of electronic products, along with EPEAT (Electronic Product Environmental Assessment Tool). In addition, as tech organizations become increasingly serious about their environmental impact and work to reduce their carbon footprints, we may see new certifications emerge that will help monitor their carbon usage while creating more energy-efficient technologies.

Collaboration and Multi-Stakeholder Initiatives

Sustainability certification is crucial for businesses to meet rapidly increasing demand. It requires collaboration between various parties, including companies, states, responsible businesses, NGOs, and accreditor bodies. Such multi-stakeholder frameworks can support greater coherence in global sustainability objectives, enhance credibility, and promote the adoption of shared standards that support the transition towards a more sustainable economy. Twenty-four governments will promote environmentally and socially responsible certification in an economically viable manner. This will mean that certification bodies work in close dialogue with regulators to ensure that new (and often evolving) standards regarding certification, such as those related to carbon and biodiversity, can be taken into account.

Certifications play a crucial role in promoting sustainability, as they demonstrate businesses' commitment to circular economy practices, carbon neutrality, and efficiency. As expectations grow, certifications will evolve to address the complexities, costs, and risks associated with greenwashing. In the coming years, sustainability certifications are expected to leverage digital technologies, enhance transparency, and foster collaboration among various stakeholders, rather than relying solely on individual companies. The future is expected to guide businesses toward a sustainable, resilient, and equitable world as the industry adjusts to the change.

CONCLUSION

Currently, sustainability certifications play a crucial role in addressing global challenges such as climate change, resource depletion, and social inequality. They provide operating capacity, a clearer business picture, and improve customer trust. These certifications also give better compliance with rules. These certifications play a vital role in fostering a more sustainable global economy, balancing social equity, environmental protection, and economic growth as the world faces increasing environmental and social challenges. Sustainability certifications play a crucial role in achieving global targets, including the United Nations Sustainable Development Goals (SDGs).

These certified products help to create a more sustainable world, protecting the balance between social, environmental, and economic needs. Sustainability certifications are essential tools for businesses to gain a competitive advantage in the face of environmental and social challenges. However, challenges, including cost, intricacy, and greenwashing, still exist. The future of sustainability certifications is set to be shaped by a combined effort among businesses, governments, NGOs, and certification bodies. This teamwork aims to create incentives that promote sustainable practices and systems. The next research may explore how to make certification accessible to small and medium-sized businesses to align with global sustainability standards. There are ways to utilize digital technology to streamline the company's existing certification processes and enhance overall sustainability. It's also about finding ways to standardize these certifications on a global level or examining their long-term effects, much like other initiatives that corporations and political systems often support. More fundamentally, we should pay attention to the invisible changes, like those affecting the environment.

REFERENCES

Amundsen, V. S. (2022). From checking boxes to actual improvement: A new take on sustainability certification. *Aquaculture (Amsterdam, Netherlands)*, *548*, 737672. DOI: 10.1016/j.aquaculture.2021.737672

Appolloni, A., Jabbour, C. J. C., D'Adamo, I., Gastaldi, M., & Settembre-Blundo, D. (2022). Green recovery in the mature manufacturing industry: The role of the green-circular premium and sustainability certification in innovative efforts. *Ecological Economics*, *193*, 107311. DOI: 10.1016/j.ecolecon.2021.107311

Borsacchi, L. (2025). Certification Standards and Ecolabelling for Sustainability and Traceability in the Textile Supply Chain. In Sustainable Digital Marketing for Fashion and Luxury Brands: Theory and Practice (pp. 675-697). Cham: Springer Nature Switzerland. DOI: 10.1007/978-3-031-82467-8_28

Daddi, T., Iraldo, F., & Testa, F. (2015). *Environmental Certification for Organisations and Products: Management approaches and operational tools*. Routledge. DOI: 10.4324/9781315768182

Dharmarathna, D., Graham, P., & Bunster, V. (2025). Unpacking the evolution of Green Star rating tools and their certification trends in Australian building projects. *Journal of Building Engineering*, *113328*, 113328. Advance online publication. DOI: 10.1016/j.jobe.2025.113328

Fotis, G., Gupta, S., Varshney, T., Dhar, S. L., & Gulzar, M. M. (Eds.). (2025). *Modern Computing Technologies for EV Efficiency and Sustainable Energy Integration*. IGI Global. DOI: 10.4018/979-8-3373-2382-4

Griffiths, K., Boyle, C., & Henning, T. F. (2018). Beyond the certification badge—How infrastructure sustainability rating tools impact on individual, organizational, and industry practice. *Sustainability (Basel)*, *10*(4), 1038. DOI: 10.3390/su10041038

Gupta, S. (2025). *The Future of Machine Learning with Automation in Engineering*. Nova Science Publishers. DOI: 10.52305/YFIX2832

Gupta, S., & Tripathi, R. K. (2014). Optimal LQR controller in CSC based STATCOM using GA and PSO. *Archives of Electrical Engineering*, *63*(3), 469–487. Advance online publication. DOI: 10.2478/aee-2014-0034

Gupta, S., & Tripathi, R. K. (2015). Transient Stability Enhancement of Multimachine Power System Using Robust and Novel Controller Based CSC-STATCOM. *Advances in Power Electronics*, *2015*(1), 626731. DOI: 10.1155/2015/626731

Heras-Saizarbitoria, I., Arana, G., & Boiral, O. (2015). Exploring the dissemination of environmental certifications in high and low polluting industries. *Journal of Cleaner Production, 89*, 50–58. DOI: 10.1016/j.jclepro.2014.10.088

Jamoussi, B., Abu-Rizaiza, A., & AL-Haij, A. (2022). Sustainable building standards, codes and certification systems: The status quo and future directions in Saudi Arabia. *Sustainability (Basel), 14*(16), 10314. DOI: 10.3390/su141610314

Johri, S., Selvakumar, P., Satyanarayana, P., Das, A., Gupta, S., & Bhattacharya, S. (2025). Urbanization and Energy: The Nexus of Global Challenges. In Geopolitical Landscapes of Renewable Energy and Urban Growth (pp. 453-478). IGI Global Scientific Publishing. DOI: 10.4018/979-8-3693-8814-3.ch016

Klooster, D. (2006). Environmental certification of forests in Mexico: The political ecology of a nongovernmental market intervention. *Annals of the Association of American Geographers, 96*(3), 541–565. DOI: 10.1111/j.1467-8306.2006.00705.x

Kuiper, G., & Gemählich, A. (2017). Sustainability and depoliticisation: Certifications in the cut-flower industry at Lake Naivasha, Kenya. *Africa Spectrum, 52*(3), 31–53. DOI: 10.1177/000203971705200302

Lima, L., Trindade, E., Alencar, L., Alencar, M., & Silva, L. (2021). Sustainability in the construction industry: A systematic review of the literature. *Journal of Cleaner Production, 289*, 125730. DOI: 10.1016/j.jclepro.2020.125730

Marchi, A. (2022). Environmental sustainability certifications in the Fashion Industry.

Poponi, S., Colantoni, A., Cividino, S. R., & Mosconi, E. M. (2019). The stakeholders' perspective within the B corp certification for a circular approach. *Sustainability (Basel), 11*(6), 1584. DOI: 10.3390/su11061584

Prell, M., Zanini, M. T., Caldieraro, F., & Migueles, C. (2020). Sustainability certifications and product preference. *Marketing Intelligence & Planning, 38*(7), 893–906. DOI: 10.1108/MIP-12-2019-0616

Rodríguez-García, R., Ferrero-Ferrero, I., & Fernández-Izquierdo, M. Á. (2023). Analysis of integration of sustainability in sustainability certifications in the hotel industry. *Frontiers in Sustainability, 4*, 1116359. DOI: 10.3389/frsus.2023.1116359

Sharma, R. K., Saran, V. B., Kanaujia, S., & Gupta, S. (2024, May). Evaluation of mppt controller performance using fuzzy logic design with wireless sensor node. In *2024 International Conference on Smart Systems for applications in Electrical Sciences (ICSSES)* (pp. 1-6). IEEE. DOI: 10.1109/ICSSES62373.2024.10561365

Shukla, R. D., Tripathi, R. K., & Gupta, S. (2010, November). Power electronics applications in wind energy conversion system: A review. In *2010 International Conference on Power, Control and Embedded Systems* (pp. 1-6). IEEE. DOI: 10.1109/ICPCES.2010.5698663

Sinha, R., Kaur, N., Gupta, S., & Thakur, P. (2023, November). Diagnosis of Parkinson's Disease using Hybrid Ensemble Technique. In *2023 International Conference on Ambient Intelligence, Knowledge Informatics and Industrial Electronics (AIKIIE)* (pp. 1-5). IEEE. DOI: 10.1109/AIKIIE60097.2023.10390458

Sinha, R., Sinha, K. K., Patel, M., Gupta, S., & Priya, S. (2024, July). Detection of leukemia disease using convolutional neural network. In 2024 5th International Conference on Image Processing and Capsule Networks (ICIPCN) (pp. 451-456). IEEE. DOI: 10.1109/ICIPCN63822.2024.00080

Tyagi, M., Gupta, S., & Khan, M. I. (2024, May). Reduction of voltage instability in complex power networks through the utilization of D-STATCOM. In 2024 4th International Conference on Advance Computing and Innovative Technologies in Engineering (ICACITE) (pp. 1662-1666). IEEE. DOI: 10.1109/ICACITE60783.2024.10616856

Velaoras, K., Menegaki, A. N., Polyzos, S., & Gotzamani, K. (2025). The role of environmental certification in the hospitality industry: Assessing sustainability, consumer preferences, and the economic impact. *Sustainability (Basel)*, *17*(2), 650. DOI: 10.3390/su17020650

Chapter 4

Decoding the Drive to Sustainability: A Customer-Centric Analysis of Sustainable Vehicle Adoption in a Key European Market

Hugo Sánchez Nieto
https://orcid.org/0009-0005-7984-0583
Universidad de Almería, Spain

Nieves García de Frutos
https://orcid.org/0000-0002-5061-0304
Universidad de Almería, Spain

ABSTRACT

The study delves on the antecedents of sustainable vehicle adoption (electric and hybrid cars) in Spain. Employing the Comprehensive Action Determination Model (CADM) as a theoretical framework, it accounts for four types of psychosocial factors, such as intentional variables, normative influences, situational variables, and habitual behaviour. Additionally, sociodemographic factors and pro-environmental behaviours considered to obtain a more complete profile of the owner. After analysing the data from 358 individuals, the results demonstrate that the CADM model effectively explains the adoption of sustainable cars, indicating that intention and perceived behavioural control are the primary determinants of sustainable car adoption. Additionally, significant differences among (non)owners are observed across sociodemographic, psychographic, and behavioural factors. Implications and practical recommendations are provided, emphasising the necessity for policies and initiatives to address psychosocial barriers and facilitate transition towards environmentally sustainable mobility.

DOI: 10.4018/979-8-3373-4495-9.ch004

1. INTRODUCTION

The transition towards sustainable mobility constitutes one of the most significant industrial transformations of our era. The urgency of this shift is emphasised by international organisations, which assert that prompt action is necessary (IEA, 2024; UNEP, 2024). Nevertheless, despite the urgency of changes, widespread adoption is not guaranteed. Ultimately, it depends on a comprehensive understanding of consumer behaviour, challenging organisations to apply Drucker's (1954) maxim—the purpose of a business is to create a customer—by leading this paradigm shift through a genuinely customer-centric approach.

This challenge is best understood through the lens of Moore's (1991) "chasm", the gap between early adopters and the mainstream market. To effectively bridge this gap, it is essential to first perform a comprehensive analysis of the consumers who have already transitioned. Who are these individuals, and what drives their decisions? Are their choices solely rational, indicative of their ecological identity, or a synthesis of both factors? Moreover, does this commitment impact other facets of their lives?

This chapter aims to address these inquiries through a bifurcated empirical examination. Initially, a comprehensive psychological framework, the Comprehensive Action Determination Model (CADM) (Klöckner & Blöbaum, 2010), will be validated employing Structural Equation Modelling (SEM) to establish a robust theoretical foundation by aggregating essential psychological determinants—such as social norms, perceived behavioural control, and intention—that influence the decision-making process of acquiring a sustainable car in the Spanish market.

Secondly, the chapter will focus on its primary objective: to construct a multi-dimensional profile of the current adopter. To this end, a binary logistic regression will be conducted on sustainable car ownership to determine the sociodemographic profile of the owner. Subsequently, multivariate analyses of variance (MANOVA) will be performed to compare the expanded behavioural profile based on Stern's typology (1999) between owners and non-owners of sustainable vehicles. The overarching aim is to extend beyond predictive analysis to provide a comprehensive characterisation that elucidates who adopts sustainable cars and the reasons behind their actions, exploring the synergies and potential conflicts—such as behavioural consistency and moral licensing—that characterise the Spanish sustainable consumer.

To guide this research, the following questions are posed:

- RQ1 To what extent does the CADM model offer a valid representation of the psychological architecture underpinning the purchase decision of a sustainable car?

- RQ2 Who are the current owners of sustainable cars in sociodemographic terms?
- RQ3 Are there differences in psychological profiles and pro-environmental behaviours between owners and non-owners of sustainable cars?

2. THEORETICAL FRAMEWORK

2.1. Definition and Characteristics of Sustainable Cars

In this chapter, a sustainable car is used as an overarching term to include all cars powered by electricity, either as their primary fuel source or to improve the efficiency of traditional vehicles, therefore, including both electric and hybrid models. On the other hand, vehicles that operate exclusively with internal combustion engines, whether diesel or petrol, are considered non-sustainable. The rationale for this classification lies in the fact that electric and hybrid automobiles possess significant environmental benefits compared to traditional vehicles (Mitropoulos et al., 2017; Tansini et al., 2022).

Electric cars do not emit greenhouse gases *per se,* and hybrids have significantly lower emission levels and greater fuel efficiency compared to their conventional counterparts (Fontaras et al., 2008). This disparity is even more evident in urban environments, where hybrids demonstrate comparatively better fuel consumption and emission reductions than in other settings (Demirdöven & Deutch, 2004). Regarding Spain, considering that 81% of the population resides in urban areas (World Bank Open Data, 2018) and that their typical journeys are predominantly metropolitan, hybrid vehicles represent a comparatively more sustainable alternative than in other nations.

Despite this, there are significant discrepancies regarding whether these vehicles are genuinely sustainable, as this pertains to addressing environmental pressures throughout each stage of the lifecycle, including design, manufacturing, usage, and disposal (Orsato & Wells, 2007). The most contentious aspects of this cycle are, on one hand, the energy mix employed to generate the electricity consumed by the vehicle, and, on the other hand, the pollution generated during raw material extraction, metal depletion, pollution associated with battery manufacturing and assembly, and its recyclability at the end of its service life (Hawkins et al., 2012). Given that these factors are particularly pertinent in the context of electric vehicles, due to the larger size of their batteries and their exclusive reliance on electricity, this discussion will focus on electric cars.

Concerning the primary factor, it has been established that electric vehicles utilising low-emission electricity sources, such as renewable energy, possess significant potential to mitigate greenhouse gas emissions throughout their life cycle (IPCC, 2023). Prior to analysing the data, it is pertinent to note that substantial enhancements are anticipated in this area, as the European Union has concurred on more stringent legislation aimed at augmenting its renewable energy capacity, thereby elevating its binding target for 2030 to no less than 42.5% (European Commission, 2023).

Proceeding to the second issue, in order to facilitate informed judgments, the *Life Cycle Assessment (LCA)* of the electric vehicle must be considered. This assessment encompasses all processes, from the extraction of raw materials to waste management at the end of the vehicle's useful life. It is estimated that electric vehicles emit approximately half the amount of CO_2 compared to conventional internal combustion engine vehicles over their entire lifespan. Approximately 70% of the environmental impact of electric vehicles is attributable to the electricity production necessary for their operation, consistent with the previous point. The lithium battery alone accounts for 15% of this impact, and this figure would likely be lower if the electricity used in manufacturing the battery originated from renewable energy sources (Messagie, 2014). Moreover, as battery recycling becomes more widespread, which is anticipated to commence around 2030 (Statista, 2023), the need for the extraction of new raw materials will significantly decrease. Each new battery produced under these conditions will be associated with reduced environmental pollution and lower emissions (Dewulf et al., 2010; Koroma et al., 2022).

2.2. Sustainable Car Adoption

The primary pro-environmental behaviour to be examined herein is the act of purchasing a sustainable vehicle, owing to its significant importance. As previously mentioned, adopting an electric or hybrid car helps to diminish the prevalence of combustion engine vehicles in circulation, thereby reducing their emissions (Bleijenberg et al., 2013).

Regarding the "who", it is essential to emphasise the significance of sociodemographic variables in the adoption of electric vehicles, as they extend beyond mere profile descriptors. Recent research in emerging markets (Alshurideh et al., 2025) demonstrates that factors such as age, income, and education serve as crucial moderators, either strengthening or weakening the influence of economic and psychological factors on the ultimate decision. Consequently, the sociodemographic characterisation of adopters is not merely a segmentation exercise; it is an indispensable and complementary step in behavioural studies aimed at understanding the heterogeneity of decision-making mechanisms within the population.

Regarding psychographic factors, pro-environmental behaviour does not originate from a singular factor, such as a positive attitude. Still, it is rather the outcome of a complex interplay of personal, social, and, critically, contextual factors that may either facilitate or hinder action (Steg & Vlek, 2009). Two primary theoretical frameworks have predominantly governed the investigation of pro-environmental behaviour. Ajzen's (1991) Theory of Planned Behaviour (TPB) posits a rational individual whose intention to act is grounded in their attitude, subjective norm (social pressure), and perceived behavioural control (PBC) (perceived ease). Conversely, Schwartz's Norm Activation Model (NAM) (1977) emphasises morality, asserting that action stems from personal norms (PN) activated by awareness of consequences (AC) and attribution of responsibility (AR). Both models have demonstrated significant predictive capability concerning pro-environmental behaviours (Bamberg & Möser, 2007), encapsulating, respectively, the pragmatic and altruistic dimensions of decision-making. Nevertheless, their explanatory capacity is constrained by two pivotal omissions that are fundamental to daily conduct: the influence of contextual factors and the strength of routine.

More recently, the Behavioural Reasoning Theory (Westaby, 2005) has gained traction to study pro-environmental behaviours that require a strong deliberative process, such as the purchase of electric cars (Uddin et al., 2024). The BRT elaborates on previous expectancy value models by including specific motives, that complement global motives such as values, attitudes and intentions (Westaby, 2005), thus providing a more complete understanding of the cognitive process.

All the previous models conflate complex situational influences with mere subjective perception overlooking the significance of objective barriers and facilitators within the environment—such as available infrastructure, actual costs, or regulations—which can entirely supersede even the strongest intentions (Tanner, 1999). Moreover, the models emphasise deliberate and conscious processes, thereby underestimating the influence of habitual behaviour. Much of human conduct, including that associated with transportation or consumption, is automatic and dictated by well-established routines that function with minimal reflection (Verplanken & Aarts, 1999). In familiar circumstances, habit serves as a shortcut that supersedes the decision-making process, rendering past behaviour a more reliable predictor of future actions than mere intention.

These limitations indicate that, for a comprehensive understanding, a model must extend beyond deliberation and morality, explicitly incorporating both objective situational constraints and the influence of automatic behaviours.

2.3. The CADM Model

The *Comprehensive Action Determination Model (*CADM), (Klöckner & Blöbaum's, 2010) may overcome such problems by proposing a framework that integrates normative, situational, habitual processes, and intentional influences as antecedents of pro-environmental behaviour. This model is founded upon the premises of Ajzen's

Theory of Planned Behaviour (TPB) (1991), Schwartz's Norm-Activation Model (NAM) (1977), Tanner's Ipsative Theory of Behaviour (1999), and the theoretical construct of habit. The incorporation of the habit component within this model addresses and overcomes the limitations identified in both the TPB and the NAM (Verplanken et al., 1994; Klöckner et al., 2003).

The application of CADM showed noteworthy results, that were able to account for a significantly greater total variance compared to other frameworks (Klöckner and Blöbaum's, 2010). This represents a substantial advancement towards developing a comprehensive theoretical framework capable of being applied to all behavioural contexts by encompassing all relevant factors that influence conduct. Given that other research involving habit measures has also demonstrated their impact on the intention to adopt electric vehicles (Moons & De Pelsmacker, 2015), the selection of the CADM as theoretical framework is expected to provide a more robust understanding of the determinants of sustainable car adoption.

The integration of all the fundamental predictors of behaviour into a single model facilitates the inclusion of all significant aspects when designing future strategies (Klöckner & Blöbaum, 2010). According to the CADM model, in this study, pro-environmental behaviour is represented by the selection of a sustainable vehicle, which is determined by four distinct types of processes. (1) *Habitual processes*, represent the repeated engagement on automatic routines that involve little cognitive effort (Verplanken & Orbell, 2003) such as the selection of transportation methods typically employed. (2) *Intentional processes* represent the result in form of willingness to act of the evaluation of the action and are the most important antecedent of action (Ajzen, 1991). In this study, intentional processes encompass the deliberate will to drive a sustainable vehicle (3) *Situational influences*, such as the perceived control regarding the usage of a sustainable car across various scenarios and car access, represent the current ability to use a sustainable car. (4) *Normative processes*, encompassing personal norms and its three antecedents: awareness of needs, awareness of consequences and social norms. Beyond this, these four factors are also related. Perceived behavioural control acts as an antecedent of both personal norms and intention. In turn, personal norms also influence intention. All these relationships are depicted on figure 1.

Figure 1. Theoretical Framework CADM Adaptation

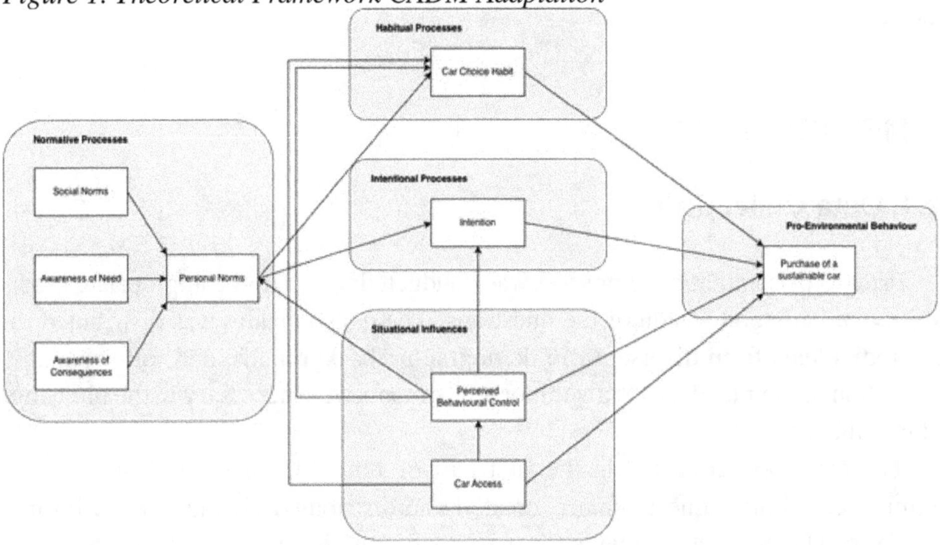

2.4. Sustainable Car Adoption and Engagement on Other Pro-Environmental Behaviours

The impact of a pro-environmental behaviour, such as sustainable car adoption—characterised by high involvement and signalling—has the potential to influence engagement with further pro-environmental behaviours. However, it is not clear to what extent this effect will be positive or negative. The existing literature presents two contrasting hypotheses to elucidate this phenomenon. On one hand, Self-Perception Theory (Bem, 1972) suggests a positive *spillover*: engaging in a notable act, such as purchasing a sustainable vehicle, reinforces the individual's environmental identity and encourages them to harmonise other actions with this self-perception to uphold consistency. Conversely, the moral licence (Mazar & Zhong, 2010) forecasts an opposing outcome. It posits that a virtuous act bestows moral credibility, which may rationalise reduced engagement in subsequent actions, as the individual perceives that they have "already done their part".

Rather than treating these theories as mutually exclusive, this study will explore whether both can coexist, with their predominance depending on the level of effort required. To this end, Stern's (1999) typology of pro-environmental behaviour will be adopted, which classifies behaviours as: green purchase behaviour (economic/informational cost), civic behaviour (low individual cost) and activist behaviour (high commitment). This structure will allow us to delve deeper into the consumer and how the purchase of a sustainable car relates to other less effortful pro-environmental behaviours.

3. METHODOLOGY

3.1. Data Collection

Prior to dissemination, a pre-test was conducted to assess the appropriate readability and comprehension of the questions. The questionnaire was distributed to ten individuals from diverse socio-demographic backgrounds, and following the collection of their feedback, only minor adjustments were necessary in the phrasing of certain questions.

The data were collected in the first half of June 2024 using a 30-item self-administered online questionnaire, created and distributed via the Google Forms platform. The only requirements for participating in the survey were residence in Spain (due to the geographical scope of the study) and the existence of at least one car in the household.

To recruit participants, a single strategy of non-probabilistic sampling by voluntary response was employed. The survey link was disseminated through multiple digital channels to maximise its reach. These included: public online forums of mobility; posts in various thematic communities on the Reddit platform; and posts on personal social media profiles such as X (formerly Twitter). Due to the low number of participants with sustainable cars, a post was uploaded featuring a prominent Spanish influencer in mobility, who had over 32,000 followers, and requesting assistance in spreading the word in order to achieve a sufficient number of respondents who were owners of sustainable cars. Following the publication of an encouraging post by the influencer urging his followers to participate in the survey, the tweet garnered 5,713 views by the conclusion of data collection.

The anonymity and confidentiality of the responses were guaranteed. Participation was voluntary, and no financial compensation was offered. The final sample was comprised of 358 responses, with 229 from owners of non-sustainable cars and 129 from owners of sustainable cars (24 hybrids and 105 electric cars).

A preliminary analysis of the data shows that our sample mainly consists of owners of non-sustainable cars (64%), and currently married individuals (36.3%). They are aged between 25 and 34 years (25.7%), possess a university education (46.1%), live in urban areas (75.7%), have a household income between €20,000 and €39,999 (34.4%), and are employed (73.2%). The gender distribution shows an important bias with an overrepresentation of men (75.70%) This may be attributable to the communities where the questionnaire was distributed, where most of the members are male and warns about a potential limitation of findings' generalization.

3.2. Measurement

The questionnaire was divided into two sections: the first section included nine demographic inquiries; the second section consisted of questions assessed on a five-point Likert scale, where the negative end of the scale was labelled "strongly disagree" (1) and the positive end as "strongly agree" (5). Each factor of the CADM components was assessed using two or three items, except for car access and pro-environmental behaviour, which were measured by only a single item. Additionally, pro-environmental behaviours based on Stern's (1999) were considered separately and evaluated through a single question or item. Detailed information is provided in the appendix, with this preliminary clarification to facilitate understanding. The data collected was coded and processed using IBM SPSS version 30 and IBM SPSS Amos version 26 statistical software.

4. ANALYSIS AND RESULTS

4.1. Refinement and Validation of the Measurement Model

The initial objective was to validate the CADM model within the studied context, executed in two phases in accordance with established best practices (Anderson & Gerbing, 1988). A fundamental preliminary step involved the recoding of car types, initially categorised into three groups: combustion, hybrid, and electric. Subsequently, the Pro-Environmental Behaviour variable was dichotomised into "Sustainable car (including hybrid and electric)" and "Non-sustainable car".

Following the model evaluation, three items were omitted due to low factor loadings or cross-loadings (see the appendix). The reliability and convergent validity of all scales in the measurement model were evaluated. The former was examined using Cronbach's alpha for internal consistency and composite reliability (CR), while convergent validity was assessed using the average variance extracted (AVE). Most constructs demonstrated robust psychometric properties. Due to inadequate reliability ($\alpha = .51$) and an AVE below 0.4, of the general habit construct, the response frequency variable (HAB1) was selected as the most appropriate proxy for such construct. Finally, the refined measurement model exhibited a good fit to the data (CFI= .969, TLI= .951, RMSEA= .051).

Discriminant validity was corroborated through the Fornell-Larcker criterion, as the square root of AVE surpassed inter-construct correlations. This indicates that each construct shares greater variance with its respective indicators than with other constructs, thus confirming strong discriminant validity.

Table 1. Discriminant validity results.

	AN	AC	SN	PN	PBC	INT
AN	**0.847**					
AC	0.746	**0.894**				
SN	0.464	0.506	**0.652**			
PN	0.556	0.513	0.423	**0.804**		
PBC	0.269	0.312	0.201	0.112	**0.734**	
INT	0.536	0.527	0.555	0.303	0.474	**0.868**

Note AN: Awareness of need, AC Awareness of consequences, SN social norm, PN personal norm, PBC perceived behavioural control, INT intention

4.2. Testing the Structural Model

Subsequently, the final structural model of the CADM demonstrated good fit to the data (CFI = .936, TLI = .914, RMSEA = .068), accounting for 40.8% of the variance in sustainable car adoption behaviour. The examination of the structural paths uncovered a decision architecture characterised by significant nuances.

The ultimate behaviour is directly and strongly predicted by PBC (β = .428, $p < .001$) and intention (β = .341, $p < .001$). Notably, habit did not manifest as a significant direct predictor ($p = .586$). In turn, intention is determined by two main forces: the influence of social norms (β = .669, $p < .001$) and a component of PBC (β = .336, $p < .001$). Notably, there is no direct effect of personal norms on intention ($p = .643$), suggesting that the motivation to purchase in the general population is based more on social validation and perceived feasibility than on internal moral conviction.

Finally, although habit does not directly affect purchasing, it is a relevant construct whose formation is positively related to situational influence (car access, β = .134, $p = .007$) and, more powerfully, inhibited by personal norms (β = -.339, $p < .001$), which acts as a resistance mechanism against automatic behaviour. The latter is influenced by both awareness of need (β = .340, $p < .001$) and awareness of consequences (β = .202, $p < .041$), with social norms not emerging as significant ($p = .281$).

Figure 2. SEM Results

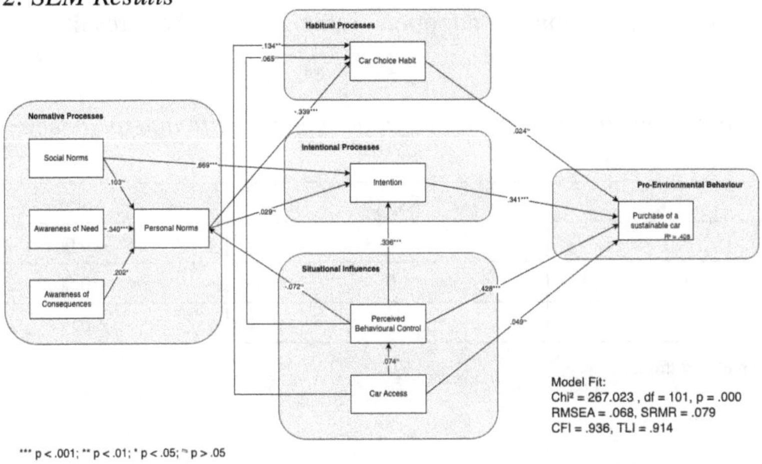

Source: own elaboration. Displayed values are standardised path coefficients.

4.3. Development of the Segmentation Model

After validating the CADM psychological model to understand the "why" of the decision, this section focuses on answering the "who" question. The objective is to develop a model based on observable sociodemographic characteristics to construct a clear profile of the consumer who leads the adoption of sustainable vehicles.

For this purpose, binary logistic regression was chosen. As predictors, sociodemographic variables gender, age range, level of education and income level were included. The final model demonstrated substantial predictive power. The Omnibus test confirmed the overall significance of the model ($\chi^2(8) = 110.757$, $p < .001$), indicating that the set of predictors is significantly better than a null model. The model explained 36.5% of the variance in sustainable car ownership, according to Nagelkerke's R^2, and achieved an overall classification accuracy of 78.2%.

The findings revealed a clear hierarchy in the importance of the predictors, allowing us to outline the profile of the adopter. Gender emerged as a highly significant predictor. The odds ratio of 6.058 (95% CI: 2.770 - 13.247) indicates that, holding other factors constant, men are more likely to own a sustainable car than women. This finding points to a notable gender gap in the early adoption of this technology. Age also proved to be a significant predictor. The odds ratio of 1.963 (95% CI: 1.579 - 2.442) suggests that increases in age range improve the odds of owning a sustainable car. The income level manifested itself as a key factor. The odds ratio of 1.513 (95% CI: 1.202 - 1.903) indicates that increments in annual income affect the probability of sustainable car adoption. Table 2 shows these results.

Table 2. Binary logistic regression coefficients for significant sociodemographic variables

Factors	Exp (B)	Sig.
Gender	6.058	< 0.001
Age	1,963	< 0.001
Income	1,513	< 0.001

Source: own elaboration.

4.4. Differences Between Groups for Pro-Environmental Behavior

Having validated the predictive models, the research focus shifts to a more in-depth characterisation of consumers who have already adopted a sustainable vehicle, allowing for a deeper exploration of the behavioural profiles of adopters versus non-adopters. To this end, a group comparison strategy was used, employing the Pro-Enviromental Behaviour variable as a segmentation factor. MANOVA was performed to compare the two groups regarding the three behaviours outlined in Stern's (1999) typology.

A noteworthy multivariate effect was observed (Pillai's Trace = .028, F(3, 354) = 3.376, p = .019), confirming that individuals who own sustainable vehicles demonstrate a distinct overall pattern of pro-environmental behaviour. Nevertheless, subsequent univariate analyses indicated a more nuanced dynamic. As shown in table 4, the difference was statistically significant for civic behaviour (e.g., recycling) (F(1, 356) = 9.936, p = .002), with owners exhibiting higher mean scores. Conversely, no statistically significant differences were observed for activist behaviour (p = .267) and only marginal effects were found for green purchase behaviour (p = .083).

Table 3. Results of the comparison of means between groups.

Factors	Type of car owned	Mean	Sig.
Civic behaviour	Non-sustainable	4.262	0.002
	Sustainable	4.605	
Activist behaviour	Non-sustainable	2.319	0.267
	Sustainable	2.473	

Factors	Type of car owned	Mean	Sig.
Green purchase behaviour	Non-sustainable	3.100	0.083
	Sustainable	3.349	

Source: own elaboration based on the Tables Tests of Between-Subjects Effects and Estimated Marginal Means

5. DISCUSSION

5.1. Applicability of CADM to Sustainable Car Adoption Process

One of the primary objectives of this research was to evaluate the generalizability of the Comprehensive Action Determination Model (CADM) within a domain of pro-environmental behaviour distinct from that in which it was originally validated. While the foundational work by Klöckner and Blöbaum (2010) concentrated on a high-frequency, low-cost decision—specifically, the selection of daily transportation modes—this study applied the model to a low-frequency, high-cost, high-involvement decision: the adoption of a sustainable vehicle. A comparison between the findings of this study and those of the original research underscores both similarities that support the validity of the overall model structure and significant differences that contribute meaningfully to the understanding of the theory.

The results corroborate several foundational aspects of the CADM, thereby affirming its validity as a comprehensive framework. Firstly, the mediating role of intention is consistently replicated, serving as the primary link between psychological predictors and subsequent actions. Secondly, the pervasive significance of PBC is verified, appearing in both studies as a robust predictor of both intention and direct behaviour, thus emphasising its function as an effective catalyst for action.

The most notable similarity is the role of personal norms as an inhibitor of habit. In both this model and that of Klöckner and Blöbaum (2010), internal moral conviction acts as a significant brake against the formation of unsustainable automatic routines. This finding suggests the existence of a universal psychological mechanism, where conscience constantly struggles against the inertia of habitual behaviour.

Beyond the similarities, the analysis uncovers three fundamental distinctions that nuance and refine the CADM theory. The most notable divergence is observed in the antecedents of intention. While Klöckner and Blöbaum (2010) determined that personal norms served as a stronger predictor than social norms in the context of choosing daily transportation, the current study reveals a markedly contrasting pattern. Specifically, in the decision-making process regarding car purchase, social norms emerge as the predominant predictor, whereas personal norms exhibit no significant direct influence. This finding implies that the focus of deliberation varies in accordance with the visibility and social relevance of the action. For private, routine decisions, individuals tend to be influenced by their moral considerations. Conversely, in high-cost, publicly observable, and identity-related decisions, affirmation from the social environment assumes primary importance.

The second significant contribution of this research involves the redefinition of the role attributed to habit. In the original framework, habit is considered a direct and influential predictor of behaviour. Conversely, within this study, its direct influence on the final action is no longer statistically significant. This observation contrasts with specific literature on pro-environmental behaviour but aligns well with the nature of the decision examined. The acquisition of a sustainable vehicle constitutes a high-cost, infrequent decision. This may have mitigated the influence of habits within this study, which become less important when individuals engage on intentional deliberative processes (Ajzen, 1991) that limit the impact of automatic routines (Verplanken & Orbell, 2003). The act of purchasing a vehicle is, primarily, characterised as an inherently rational and deliberate decision, wherein habitual automaticity is effectively disengaged. Therefore, habit may rather exhibit greater influence in contexts characterised by low deliberation yet become attenuated by conscious processes such as intention and PBC.

The CADM structural model indicates that owning a sustainable vehicle is directly influenced by both intention and PBC, with the latter having the most significant impact. This result suggests that, at the final moment of decision-making, the balance is tipped not so much by "wanting" (intention) as by perceived "ability" (PBC). Our analysis further indicates the dual function of the PBC within the decision-making framework. It serves not only as the primary direct predictor of behaviour but also plays an integral role in the formation of intention. Consumer self-efficacy, their confidence that they can overcome practical barriers (burden, autonomy), is the key that opens the door to intention and, ultimately, to action. This finding is corroborated by extant research. Using a multi-criteria decision model with a panel of experts, Seikh and Chatterjee (2025) concluded that economic (price) and technical (range, recharging) challenges—both evaluated under of PBC—are the most critical obstacles to mass adoption. This finding is consistent across various methodologies. Notably, Higueras-Castillo et al. (2020), employing a machine learning approach, arrived at a similar conclusion: autonomy and reliability—core components of the PBC—are, together with incentives, the most significant factors in predicting purchase intent. This demonstrates that perceived barriers exert a compounded negative impact: initially by preventing the development of strong intentions and subsequently by obstructing the translation of existing intentions into action.

Furthermore, corroborating the central tenets of CADM, intention functions as a direct predictor of action. This discovery emphasises the deliberate nature of choosing to adopt a sustainable vehicle. Such behaviour is neither impulsive nor solely habitual; rather, it stems from a conscious decision-making process where the development of an explicit intention serves as an essential link between normative reflection and the eventual conduct. In essence, intention represents, in this context, a potential critical failure point in the adoption process. The "*attitude-behaviour gap*" identified by Kollmuss and Agyeman (2002) appears, according to our sample, to be a gap in the formation of intention.

Although intention plays a crucial role in the execution of pro-environmental behaviour, an analysis of its antecedents reveals that it is predominantly shaped not by morality, but by pragmatism and social validation. The significant influence of social norms and PBC on intention—in contrast to the non-significant effect of personal norms—stands as a key conclusion of this research. It indicates that, in cases of high visibility purchasing decisions, consumer intention is predominantly socially constructed: it is formed through observing the actions of others and assessing the practicality of the behaviour. This conclusion aligns strongly with the most recent international research. In a cross-cultural study, Higueras-Castillo et al. (2024) found that "social value"— the vehicle's ability to enhance image and status—was the most potent predictor of purchase intention, even above price and ecological value. This reinforces our conclusion that adopting a sustainable car, as a high-visibility decision, is fundamentally an act of social construction and validation.

Regarding the significant influence of social norms on intention, it is noteworthy that this association has been documented in scholarly literature by authors such as Fishbein and Ajzen (2010) and Abrahamse and Steg (2013), who demonstrated that these norms considerably affect individuals' intentions, consistent with observations in our sample.

Ultimately, the explanatory power of the structural model is inferior to that achieved in its original application (Klöckner & Blöbaum, 2010). Forecasting the purchase of a high-cost durable item, influenced by numerous unmeasured economic and market factors, is inherently more challenging than predicting a daily transportation choice. Therefore, this should not be construed as a weakness of the model but rather as a reflection of the complexity of the phenomenon examined. The fact that a psychosocial model accounts for over 40% of the variance in such a multifaceted behaviour is compelling evidence of the utility of the CADM framework.

One possible explanation for this high value could be a consequence of the sampling carried out. Given that social norms are a person's perceptions of normal and accepted behaviour within a group (Schultz, 2022), by sending our survey to forums and communities specialising in sustainable mobility, it could have been answered by people with high social norms; these people may have made the purchase seeking social validation, which they find in forums frequented by people like them. In addition, it is worth highlighting another approach to purchasing a sustainable car, especially those that involve a greater change in daily use (pure electric cars). This purchase is both a high-cost, high-risk decision and a paradigm shift for the potential owner which translates, respectively, into a greater influence of social proof (Kahneman, 2011) and a social norm that guides behaviour (Cialdini & Goldstein, 2004).

Considering our findings on the significance of PBC and social norms, it is evident that they align with recent research utilising alternative theoretical frameworks. For instance, Uddin et al. (2024), employing BRT, identified that "motives for" adoption —such as environmental awareness and technology —and "motives against" —such as price sensitivity and scepticism —serve as key predictors of attitude and intention. The constructs within our CADM model can be interpreted as the operationalisation of these underlying reasoning processes: social norms function as a compelling "motive for", whereas barriers to PBC can be understood as "motive against".

In summary, answering RQ1, this research confirms the applicability of the CADM as a suitable model to explain sustainable car adoption, which allows extending it beyond its initial context to explain a high-effort one-time adopted pro-environmental behaviour. Furthermore, it demonstrates that the model is not a rigid framework; rather, the significance of its components is adaptable and varies according to the characteristics of the behaviour. This indicates that the psychology underlying pro-environmental decision-making undergoes a fundamental transformation as the costs and visibility associated with the action escalate.

5.2. Sociodemographic Factors Affecting Sustainable Car Ownership

The results also provided a clear answer for RQ2, showing how being male, older, and having a higher income affect the likelihood of owning a sustainable car. Although gender emerged as the most significant predictor, this finding warrants cautious interpretation due to the predominance of male respondents across the sample (just over 75%). The magnitude of the gender effect could be partially exaggerated due to the sampling procedure developed via forums dedicated to sustainable motoring.

Conversely, the remaining two factors (age and income) may possess a logical foundation and be interconnected. The findings from the sociodemographic analysis expose a tension between psychological factors and socio-economic realities. This is the juncture where many young individuals find themselves. Due to their lower-paying jobs or the absence of accumulated savings relative to older populations, they face greater financial constraints (Barbarossa & De Pelsmacker, 2014). Economic prosperity and social class are positively correlated with concern for the environment and environmental awareness (Leonidou et al., 2015), which, as previously elucidated, serve as precursors to environmental consciousness and pro-environmental conduct (Sheltzer et al., 1991). Moreover, sustainable vehicles, particularly electric automobiles, convey a higher initial purchase cost compared to internal combustion engine vehicles (Degirmenci & Breitner, 2017), with electric cars exhibiting a price differential of approximately 10% to 50% within the European and United States markets (International Energy Agency, 2024), and hybrid cars costing between €5,000 and €12,000 more (Van Vliet et al., 2010). Consequently, individuals with greater wealth have a greater ability to engage in pro-environmental behaviours, such as acquiring a sustainable car. This is attributable both to their heightened environmental awareness and their enhanced economic capacity.

It is important to note that Alshurideh et al. (2025) discovered that demographic variables influence the manner in which consumers interpret economic and psychological benefits. For instance, older consumers with lower incomes exhibit greater sensitivity to financial incentives, whereas younger and more educated consumers respond more favourably to technological innovation and environmental awareness. This implies that the demographic profile identified reflects varying sensitivities to persuasive strategies.

5.3. Differences in Pro-Environmental Behaviour Engagement Between Groups

Regarding the answer to RQ3, this study also found evidence that sustainable and non-sustainable car owners also differ in their levels of engagement with other pro-environmental behaviours. An interesting finding of this research is the nuanced pattern of behavioural *spillover* from sustainable car adoption towards other pro-environmental behaviours. The findings disclosed a complex dynamic that endorses the coexistence of the mechanisms outlined in Self-Perception Theory (Bem, 1972) and moral licence (Mazar & Zhong, 2010), functioning at various levels of a continuum of effort. Whereas a highly significant difference was observed in civic behaviour—with sustainable car owners reporting higher engagement, only marginal or no significance was obtained for green purchase behaviour and activist behaviour, respectively.

As Gleim et al. (2013) argue, the perceived efficacy of behaviour plays a pivotal role in shaping pro-environmental engagement. Low-effort actions such as recycling and maintaining environmental cleanliness (i.e., civic behaviour) are likely to provide immediate and visible results, reinforcing an individual's sense of impact and affirming their self-concept as a responsible citizen. Having established a moral self-image through simple, high-feedback behaviours, individuals may feel justified.

Yet, as behavioural costs or ambiguities increase, this sense of efficacy weakens, and psychological mechanisms such as moral licensing (Mazar & Zhong, 2010) may become more salient in relaxing subsequent efforts, since individuals interpret their prior actions as sufficient evidence of their environmental commitment.

In the domain of green purchasing, this dynamic intensifies. Beyond financial considerations, consumers often transfer responsibility to external agents—companies whose environmental claims may be met with scepticism and uncertainty (Gleim et al., 2013). Such ambiguity fosters cognitive barriers and creates fertile ground for moral licensing: the perception that one's earlier "good deed" entitles one to disengage from more demanding forms of activism. As Thøgersen and Ölander (2003) suggest, individuals frequently choose easy actions to avoid costlier commitments, a tendency echoed in the notion of a limited psychological "effort budget." As the results of this chapter support, the adoption of a sustainable vehicle, may deplete this budget, effectively legitimising the avoidance of further pro-environmental engagement on high-cost actions such as purchase or activism—a decoupling observed by Summers et al. (2016) across behavioural contexts.

In conclusion, the adopter's profile does not correspond to that of an individual exhibiting universally pro-environmental behaviour. Rather, it describes a person who, following an act of substantial commitment that redefines their identity, seeks consistency through low-cost, high-visibility civic behaviours. Nonetheless, this consistency is constrained by a threshold of effort, beyond which scepticism, complexity, and the perception of having met environmental obligations trigger a moral licence that rationalises inaction.

6. LIMITATIONS AND FUTURE RESEARCH VENUES

This study presents several limitations that must be acknowledged. In turn, such limitations show potential areas where future research can provide fruitful answers.

6.1. Cross-Cultural Differences

Firstly, the non-probabilistic sampling methods employed, along with an overrepresentation of male participants recruited, constrain the extent to which the findings can be generalised to the broader Spanish population. In particular, the significant predictive effect of gender warrants cautious interpretation. Therefore, this study requires replication with balanced and representative samples to validate the true extent of demographic effects and the organisation of psychological profiles. National cultural dimensions can significantly moderate the relative importance of predictors of electric vehicle adoption (Higueras-Castillo et al., 2023). Therefore, the findings presented here, especially the strong influence of Social Norms, may be specific to the Spanish cultural context (i.e., relatively higher collectivism) and should not be generalised to other cultures without further empirical validation.

Consequently, a promising avenue of research emerging from this study is the implementation of a cross-cultural investigation. Replicating the validated CADM model in diverse national settings, with different levels of relevant cultural dimensions would enable us to assess the universality of our findings. Following Higueras-Castillo et al. (2023), it would be highly beneficial to examine how variables such as individualism-collectivism and power distance influence the relationships among CADM constructs, thereby providing a genuinely global perspective on the psychology of sustainable mobility.

Similarly, in line with recent works (Alshurideh et al., 2025) it would be beneficial to evaluate how demographic variables moderate the effects of psychographics—i.e., perceived behavioural control or social norm—on sustainable car adoption depending on the country setting.

6.2. Longitudinal Impact of Policy and Market Conditions

Secondly, the cross-sectional design of the study provides a static snapshot of the decision-making process; however, it does not facilitate an analysis of the evolving perceptions and intentions over time. Longitudinal designs that monitor consumers over time are essential for understanding how significant events (e.g., the installation of a workplace charger, a rise in petrol prices) influence PBC and, subsequently, intention and behaviour.

This longitudinal approach is also the ideal method for testing the resilience of our findings under changing political and market conditions. For example, it would allow us to empirically analyse the impact of shock events such as an abrupt withdrawal of subsidies or the implementation of much stricter emissions laws, or disruptive trends such as autonomous driving and shared mobility functions. Such a design could reveal whether, in an environment without economic incentives or where the PBC shifts to assessing the service's suitability rather than the vehicle's ownership, the relative importance of CADM predictors is reconfigured. This type of forward-looking analysis is indispensable for developing transition strategies that are not only effective in the short term but also robust and adaptive in the long term.

The previous can be addressed by implementing a two-phase longitudinal study, drawing inspiration from the dual framework of Higueras-Castillo et al. (2025). The initial phase would replicate our current adoption model, the CADM, while the subsequent phase would involve re-surveying the same participants one or two years following their purchase of a sustainable vehicle. This approach would enable us to examine whether the predictors of continued use intention (retention) evolve, potentially shifting from normative influences (such as social norms) to purely experiential factors, including satisfaction and confirmation of expectations related to cost, range, and charging convenience. Such a study would provide a holistic understanding of the sustainable consumer's lifecycle, from initial consideration to eventual loyalty.

6.3 Consideration of Alternative Sustainable Vehicles

Another significant limitation was the aggregation of all sustainable vehicles into a single category. Future research should distinguish between hybrids (HEVs), plug-in hybrids (PHEVs), and pure electric vehicles (EVs) to evaluate the hypothesis of a “confidence cascade”, whereby PHEVs serve as an intermediary product that enhances PBC, thereby facilitating the widespread adoption of electric vehicles.

While our research has identified the important antecedents of sustainable car adoption, it is focused on EVs and HEVs, leaving aside other sustainable trends on mobility such as autonomous driving (Schwing et al., 2025) and shared mobility (Shaheen and Pan, 2024). Yet, the application of the CADM model could be considered as a fruitful venue for expanding the research on these fields. The examination of autonomous features in electric and hybrid cars—such as telematics, automated charging, or self-driving capabilities—could substantially influence consumer behaviour by simplifying navigation and vehicle management, thus enhancing perceived ease of use and usefulness, being the latter a key influencing variable of autonomous vehicles adoption (Schwing et al., 2025). Conversely, research shows how the heterogeneity of drivers plays an important role on autonomous vehicle adoption (Li et al., 2024) and therefore, for some individuals, increased automation may reduce PBC, as reliance on complex technology could generate feelings of loss of control or distrust in the vehicle’s decision-making systems. Similarly, the rise of shared mobility, introduces new social and behavioural dynamics that may reshape norms around ownership, autonomy, and environmental responsibility. These shared systems not only challenge traditional notions of personal mobility but also promote collective, sustainability-oriented practices. They are also likely to provide access to more sustainable vehicles for those segments of individuals (i.e., younger) that do not have required resources to acquire an EV (Shaheen and Pan, 2024). Future research should therefore examine how the interplay between autonomy, connectivity, and shared mobility affects both personal and shared EV adoption, recognizing that automation can simultaneously empower and disempower users depending on their technological trust and behavioural tendencies.

6.4. Theoretical Extensions

Ultimately, it should be noted that the scope of the CADM model, while explaining a substantial portion of the variance, inherently excludes the direct influence of exogenous variables with a significant impact on behaviour—such as economic incentives, subsidies, or tax exemptions—that can facilitate the adoption of green technologies (Carlucci et al., 2018), independently of the consumer's psychological profile.

At the model level, suboptimal psychometric properties for the habit construct were obtained, resulting on its operationalisation through a single indicator. This underscores the ongoing challenge within the scholarly literature to quantify automatic behaviours reliably. A promising avenue for future research involves the integration of the CADM with the BRT. While our study has validated the causal structure of the CADM, a subsequent model could investigate how the "pro" and "con" motives identified by Uddin et al. (2024) serve as mediators between more general values and beliefs (such as AN and AC) and the development of CADM attitudes and norms, given that controversy over the actual sustainability of a product may constitute a significant barrier (Peattie, 1999). Such a model would provide a more detailed understanding of the architecture underlying consumer decision-making.

7. PRACTICAL RECOMMENDATIONS

The findings of this research, consistent with other approaches to strategic prioritisation, suggest that the most efficacious solutions for the adoption of sustainable vehicles are fundamentally structural. For instance, Seikh and Chatterjee (2025) identified the "introduction of government policies and incentives" and the "installation of charging infrastructure" as the two most influential strategies. Our results further refine this perspective by elucidating the psychological mechanisms underlying these strategies: the development of PBC and social norms.

Consequently, the subsequent recommendations are aimed at the two primary stakeholders capable of implementing these structural modifications: the private sector (manufacturers and distributors) and the public sector (regulators and administrations). The fundamental principle underpinning these recommendations is that progress depends on a synergistic action, moving beyond what Porter and van der Linde (1995) termed the "stalemate" between regulators and industry. Instead of viewing environmental goals as a zero-sum game, an effective strategy requires a collaborative effort. This cooperative strategy must, on one hand, address the market through segmentation, and on the other hand, focus on systematically removing obstacles that hinder PBC, which has become the central element of adoption and is vital in facilitating the other key factor, intention.

7.1. For Practitioners

The initial strategic recommendation is to prioritise the development of the "*Whole Product*" as defined by Moore (1991), aiming to dismantle barriers to PBC. The analysis conducted demonstrates that PBC is not only the most significant direct predictor of purchasing behaviour but also the psychological trait that most distinctly separates adopters from non-adopters. This suggests that the primary obstacle is not a lack of desire but rather a fear of complexity and a perception of infeasibility. Many environmentally friendly products and services are innovative and necessitate consumers to adopt new behaviours (Peattie & Crane, 2005), such as planning travel in advance based on their vehicle's range or adapting to frequent charging requirements. Consequently, investments should be directed towards eliminating both tangible and perceived frictions.

To this end, it is crucial to diminish the functional disparity against traditional cars concerning lower battery car range, ensuring that environmental sustainability does not compromise vehicle performance. This is particularly pertinent given that a limited number of consumers are inclined to pay a premium for environmentally friendly products that fail to satisfy their needs fully, despite inherent environmental advantages (Olson, 2012). Currently, notwithstanding the general increase in battery capacity and technological advancements, the weighted average range within the most popular segment—SUVs—approximates 380 km. This represents a substantial enhancement from previous years, with an increase of nearly 75% between 2015 and 2023; however, it remains inferior to that of conventional vehicles (International Energy Agency, 2024). The disparity is more pronounced in smaller vehicle segments, where smaller battery sizes exacerbate range limitations.

Conversely, providing turnkey solutions that integrate the vehicle with an ecosystem of services, such as guaranteed installation of charging points or unified access to pay-as-you-go charging networks, is a strategic approach. A fundamental strategy in this domain involves positioning plug-in hybrid vehicles (PHEVs) not solely as a product but also as a tool to foster market PBC and enhance consumer intention through it and social norms (social proof). These vehicles serve as a bridge product (Christensen & Raynor, 2013), which demystifies the charging experience and prepares consumers for future adoption of fully electric vehicles (EVs).

The second recommendation for enterprises is to implement a dual marketing strategy to facilitate the "*crossing of the chasm*" (Moore, 1991) and target not only *early adopters* (Rogers, 1962) but also the mainstream market. Accordingly, mass media campaigns should emphasise social proof by incorporating testimonials and constructing a narrative of community belonging. Conversely, marketing efforts directed at the segment that has already adopted this type of vehicle should be pragmatic, highlighting performance data, cost savings, and, most importantly, the ease with which the car integrates into their daily lives.

Third, given the higher price of sustainable vehicles, it can be inferred that consumers with limited financial resources (such as younger ones) are unable to participate in certain pro-environmental behaviours, such as the purchase of sustainable automobiles, primarily due to the limited variety of products available at reasonable prices (Young et al., 2009). This issue is particularly pronounced with electric vehicles, as, although some more affordable options exist, they tend to have markedly lower ranges compared to those offered by premium or luxury brands (PRNewswire, 2023). Consequently, producers need to focus on providing options to consumers with reduced purchasing power, who are currently left without adequate options that sufficiently address their needs.

Finally, the private sector must take the lead with transparency to combat consumer scepticism. To address this, companies are required to provide clear traceability of materials, from their origin to their destination after recycling, as well as their lifecycle carbon footprint. Similarly, the utilisation of tools such as gamification to convert driver behaviour into tangible environmental impact data can reinforce perceptions of effectiveness and the value of their actions, thereby empowering consumers (Ottman, 1998).

7.2. For the Public Sector

The initial recommendation is to formulate infrastructure policies targeted at augmenting PBC. Uncertainty concerning the charging network represents one of the predominant psychological barriers (Falchetta & Noussan, 2021, Moradloo et al., 2025). Consequently, the primary focus should be on expediting universal, dependable, and standardised coverage, addressing the disparities in growth among countries and, by extension, among their regions (International Energy Agency, 2024), while also fostering the interoperability of systems.

Secondly, incentive policies currently hold a crucial role, given that sales within the sector are highly reliant on them (Clinton & Steinberg, 2019). The absence of such policies results in a considerable rise in consumer prices and a subsequent decline in demand (BloombergNEF, 2024; Popovich, 2024). Bureaucratic complexities and the volatility associated with subsidies undermine consumer confidence and reduce their PBC, thereby hindering the adoption of sustainable vehicles. Beyond their direct economic effects, recent academic literature emphasises the psychological influence of incentives as catalysts for PBC. The research conducted by Uddin et al. (2024) demonstrates that the implementation of financial incentives functions as a moderating factor, alleviating the adverse influence of "counterarguments"—such as concerns regarding price and range—on consumers' purchase intentions. Similarly, Higueras-Castillo et al. (2019) found that incentives not only enhance attitudes but also actively reduce psychological barriers associated with high costs. These findings are entirely consistent with the conclusions of our study demonstrating that incentives are not merely price reductions; rather, they serve as a strategic tool for PBC management. Their primary role is to diminish the cognitive load experienced by consumers, rendering barriers appear less formidable and consequently enhancing their self-efficacy. A well-conceived incentive policy, therefore, not only reduces the cost of the vehicle but, more significantly, renders the decision to purchase it more manageable.

8. CONCLUSION

This research aimed to elucidate the architecture underlying the decision to purchase a sustainable vehicle, seeking to go beyond the psychological "why" and to construct an actionable profile of the "who" that drives adoption. The findings indicate that the distinctive characteristic of this adopter is not merely heightened environmental awareness but predominantly an elevated level of PBC. Fundamentally, the adopter is an individual who has surmounted the psychological barrier of feasibility, converting latent concerns into a definitive intention to act.

This finding suggests that the primary barrier to "*crossing the chasm*" into the mass market is not a deficiency in pro-environmental morality, but rather a perceived lack of capability. Consequently, the acceleration of sustainable mobility does not rely on a singular lever but rather on the intentional development of an adoption ecosystem. Corporations must extend their efforts beyond merely selling vehicles to orchestrate an experience that cultivates consumer PBC. Concurrently, the public sector must collaborate by establishing an infrastructure and incentive framework that diminishes uncertainty and promotes social norms, thereby fostering an environment in which the formation of an intention becomes the logical and natural outcome.

The challenge, therefore, extends beyond merely selling cars. It involves transitioning from a confrontational paradigm to one centred on dynamic collaboration between the public and private sectors, to foster social trust, deliberate intention, and, above all, the pragmatic capacity to translate consideration into purchase. The objective is not to identify the limited number of already "activated agents" but to develop a system that engenders them on a large scale, rendering sustainable mobility a valuable, viable, and tangible alternative for everyone.

REFERENCES

Abrahamse, W., & Steg, L. (2013). Social influence approaches to encourage resource conservation: A meta-analysis. *Global Environmental Change*, *23*(6), 1773–1785. DOI: 10.1016/j.gloenvcha.2013.07.029

Alshurideh, M. T., Alzoubi, H. M., Kurdi, B. A., Hamadneh, S., Ahmed, G., Al-Sulaiti, K., Bataineh, A. Q., Alquqa, E. K., & Ozturk, I. (2025). Consumer and economic influences on electric vehicle adoption: The mediating role of attitudes and the moderating effect of demographics. *International Journal of Energy Economics and Policy*, *15*(3), 214–229. DOI: 10.32479/ijeep.19313

Anderson, J. C., & Gerbing, D. W. (1988). Structural equation modeling in practice: A review and recommended two-step approach. *Psychological Bulletin*, *103*(3), 411–423. DOI: 10.1037/0033-2909.103.3.411

Bamberg, S., & Möser, G. (2007). Twenty years after Hines, Hungerford, and Tomera: A new meta-analysis of psycho-social determinants of pro-environmental behaviour. *Journal of Environmental Psychology*, *27*(1), 14–25. DOI: 10.1016/j.jenvp.2006.12.002

Barbarossa, C., & De Pelsmacker, P. (2014). Positive and negative antecedents of purchasing eco-friendly products: A comparison between green and non-green consumers. *Journal of Business Ethics*, *134*(2), 229–247. DOI: 10.1007/s10551-014-2425-z

Bem, D. J. (1972). Self-Perception Theory. *Advances in Experimental Social Psychology*, *6*, 1–62. DOI: 10.1016/S0065-2601(08)60024-6

Bleijenberg, A., Egenhofer, C., Behrens, A., Rizos, V., & Alessi, M. (2013). *Pathways to low carbon transport in the EU – From Possibility to reality*. Centre for European Policy Studies (CEPS). https://ssrn.com/abstract=2278809

Bloomberg, N. E. F. (2024). Electric Vehicle Outlook Report 2024. https://about.bnef.com/electric-vehicle-outlook/

Bukoff, A. (2020). Inside the Auto Industry: Widespread Skepticism about Electric Vehicles. SAE Technical Paper 2020-01-0529. DOI: 10.4271/2020-01-0529

Carlucci, F., Cirà, A., & Lanza, G. (2018). Hybrid electric vehicles: Some theoretical considerations on consumption behaviour. *Sustainability (Basel)*, *10*(4), 1302. DOI: 10.3390/su10041302

Christensen, C., & Raynor, M. (2003). *The Innovator's Solution: Creating and Sustaining Successful Growth*. Harvard Business Review Press.

Cialdini, R. B., & Goldstein, N. J. (2004). Social influence: Compliance and conformity. *Annual Review of Psychology*, *55*(1), 591–621. DOI: 10.1146/annurev.psych.55.090902.142015 PMID: 14744228

Clinton, B. C., & Steinberg, D. C. (2019). Providing the Spark: Impact of financial incentives on battery electric vehicle adoption. *Journal of Environmental Economics and Management*, *98*, 102255. DOI: 10.1016/j.jeem.2019.102255

Degirmenci, K., & Breitner, M. H. (2017). Consumer purchase intentions for electric vehicles: Is green more important than price and range? *Transportation Research Part D, Transport and Environment*, *51*, 250–260. DOI: 10.1016/j.trd.2017.01.001

Demirdöven, N., & Deutch, J. (2004). Hybrid cars now, fuel cell cars later. *Science*, *305*(5686), 974–976. DOI: 10.1126/science.1093965 PMID: 15310893

Dewulf, J., Van Der Vorst, G., Denturck, K., Van Langenhove, H., Ghyoot, W., Tytgat, J., & Vandeputte, K. (2010). Recycling rechargeable lithium ion batteries: Critical analysis of natural resource savings. *Resources, Conservation and Recycling*, *54*(4), 229–234. DOI: 10.1016/j.resconrec.2009.08.004

Drucker, P. F. (1954). *The Practice of Management*. Heinemann.

European Commission. (2023). Aceleración del despliegue de las energías renovables. *European Commission*. https://ec.europa.eu/commission/presscorner/detail/es/ip_23_2061

Falchetta, G., & Noussan, M. (2021). Electric vehicle charging network in Europe: An accessibility and deployment trends analysis. *Transportation Research Part D, Transport and Environment*, *94*, 102813. DOI: 10.1016/j.trd.2021.102813

Fishbein, M. & Ajzen, I. (2010). Predicting and changing behavior. DOI: 10.4324/9780203838020

Fontaras, G., Pistikopoulos, P., & Samaras, Z. (2008). Experimental evaluation of hybrid vehicle fuel economy and pollutant emissions over real-world simulation driving cycles. *Atmospheric Environment*, *42*(18), 4023–4035. DOI: 10.1016/j.atmosenv.2008.01.053

Gleim, M. R., Smith, J. S., Andrews, D., & Cronin, J. J.Jr. (2013). Against the Green: A multi-method examination of the barriers to green consumption. *Journal of Retailing*, *89*(1), 44–61. DOI: 10.1016/j.jretai.2012.10.001

Groening, C., Sarkis, J., & Zhu, Q. (2018). Green marketing consumer-level theory review: A compendium of applied theories and further research directions. *Journal of Cleaner Production*, *172*, 1848–1866. DOI: 10.1016/j.jclepro.2017.12.002

Hawkins, T. R., Singh, B., Majeau-Bettez, G., & Strømman, A. H. (2012). Comparative Environmental Life cycle assessment of conventional and electric vehicles. *Journal of Industrial Ecology*, *17*(1), 53–64. DOI: 10.1111/j.1530-9290.2012.00532.x

Higueras-Castillo, E., Guillén, A., Herrera, L. J., & Liébana-Cabanillas, F. (2020). Adoption of electric vehicles: Which factors are really important? *International Journal of Sustainable Transportation*, *15*(10), 799–813. DOI: 10.1080/15568318.2020.1818330

Higueras-Castillo, E., Kalinic, Z., Marinkovic, V., & Liébana-Cabanillas, F. J. (2019). A mixed analysis of perceptions of electric and hybrid vehicles. *Energy Policy*, *136*, 111076. DOI: 10.1016/j.enpol.2019.111076

Higueras-Castillo, E., Liébana-Cabanillas, F., Dos Santos, M. A., Zulauf, K., & Wagner, R. (2024). Do you believe it? Green advertising skepticism and perceived value in buying electric vehicles. *Sustainable Development (Bradford)*, *32*(5), 4671–4685. DOI: 10.1002/sd.2932

Higueras-Castillo, E., Rialp-Criado, J., Buhmann, K., & Liebana-Cabanillas, F. (2025). “Should I stay or should I go?” A Dual-Theory Behavioral Analysis of Electric vehicle adoption and Retention. *Transport Policy*, *103792*, 103792. Advance online publication. DOI: 10.1016/j.tranpol.2025.103792

Higueras-Castillo, E., Singh, V., Singh, V., & Liébana-Cabanillas, F. (2023). Factors affecting adoption intention of electric vehicle: A cross-cultural study. *Environment, Development and Sustainability*, *26*(11), 29293–29329. DOI: 10.1007/s10668-023-03865-y

International Energy Agency. (2024). *Global EV Outlook 2024*. https://www.iea.org/reports/global-ev-outlook-2024

IPCC. (2023). AR6 Synthesis Report. *IPCC*. https://www.ipcc.ch/report/sixth-assessment-report-cycle/

Kahneman, D. (2011). *Thinking, Fast and Slow*. Penguin Books.

Kalafatis, S. P., Pollard, M., East, R., & Tsogas, M. H. (1999). Green marketing and Ajzen’s theory of planned behaviour: A cross-market examination. *Journal of Consumer Marketing*, *16*(5), 441–460. DOI: 10.1108/07363769910289550

Kline, R. B. (2016). *Principles and practice of structural equation modeling* (4th ed.). The Guilford Press.

Klöckner, C. A., & Blöbaum, A. (2010). A comprehensive action determination model: Toward a broader understanding of ecological behaviour using the example of travel mode choice. *Journal of Environmental Psychology*, *30*(4), 574–586. DOI: 10.1016/j.jenvp.2010.03.001

Klöckner, C. A., & Matthies, E. (2009). Structural modeling of car use on the way to the university in different settings: Interplay of norms, habits, situational restraints, and perceived behavioral control. *Journal of Applied Social Psychology*, *39*(8), 1807–1834. DOI: 10.1111/j.1559-1816.2009.00505.x

Klöckner, C. A., Matthies, E., & Hunecke, M. (2003). Problems of operationalizing habits and integrating habits in normative Decision-Making models. *Journal of Applied Social Psychology*, *33*(2), 396–417. DOI: 10.1111/j.1559-1816.2003.tb01902.x

Kollmuss, A., & Agyeman, J. (2002). Mind the Gap: Why do people act environmentally and what are the barriers to pro-environmental behavior? *Environmental Education Research*, *8*(3), 239–260. DOI: 10.1080/13504620220145401

Koroma, M. S., Costa, D., Philippot, M., Cardellini, G., Hosen, M. S., Coosemans, T., & Messagie, M. (2022). Life cycle assessment of battery electric vehicles: Implications of future electricity mix and different battery end-of-life management. *The Science of the Total Environment*, *831*, 154859. DOI: 10.1016/j.scitotenv.2022.154859 PMID: 35358517

Leonidou, L. C., Coudounaris, D. N., Kvasova, O., & Christodoulides, P. (2015). Drivers and Outcomes of Green Tourist Attitudes and Behavior: Sociodemographic Moderating Effects. *Psychology and Marketing*, *32*(6), 635–650. DOI: 10.1002/mar.20806

Li, G., Yu, W., Chen, X., Wang, W., Guo, H., & Jiang, X. (2024). Exploring User Acceptance of Autonomous Vehicles: Impact of Driver and Vehicle Styles. *SAE International Journal of Connected and Automated Vehicles, 8*, 193-201. DOI: 10.4271/12-08-02-0015

Mazar, N., & Zhong, C. (2010). Do green products make us better people? *Psychological Science*, *21*(4), 494–498. DOI: 10.1177/0956797610363538 PMID: 20424089

Messagie, M. (2014). Life cycle analysis of the climate impact of electric vehicles. *Journal of Life Cycle Assessment*, 14. https://scholar.google.be/citations?view_op=view_citation&hl=nl&user=3qpRhxwAAAAJ&citation_for_view=3qpRhxwAAAAJ:hFOr9nPyWt4C

Mitropoulos, L. K., Prevedouros, P. D., & Kopelias, P. (2017). Total cost of ownership and externalities of conventional, hybrid and electric vehicle. *Transportation Research Procedia*, *24*, 267–274. DOI: 10.1016/j.trpro.2017.05.117

Moons, I., & De Pelsmacker, P. (2015). An extended decomposed theory of planned behaviour to predict the usage intention of the electric car: A Multi-Group Comparison. *Sustainability (Basel)*, *7*(5), 6212–6245. DOI: 10.3390/su7056212

Moore, G. A. (1991). *Crossing the Chasm: Marketing and Selling High-Tech Goods to Mainstream Customers*. Harper Business.

Moradloo, N., Mahdinia, I., & Khattak, A. (2024). Charging into the Future: Unraveling the Factors Shaping Electric Vehicle Adoption and Addressing Heterogeneity. *SAE International Journal of Sustainable Transportation, Energy, Environment, & Policy, 6*, 51-63. .DOI: 10.4271/13-06-01-0005

Olson, E. (2012). It's not easy being green: The effects of attribute tradeoffs on green product preference and choice. *Journal of the Academy of Marketing Science*, *41*(2), 171–184. DOI: 10.1007/s11747-012-0305-6

Orsato, R., & Wells, P. (2007). The Automobile Industry & Sustainability. *Journal of Cleaner Production*, *15*(11–12), 989–993. DOI: 10.1016/j.jclepro.2006.05.035

Ottman, J. A. (1998). *Green marketing: opportunity for innovation*. NTC Business Books.

Peattie, K. (1999). Trappings versus substance in the greening of marketing planning. *Journal of Strategic Marketing*, *7*(2), 131–148. DOI: 10.1080/096525499346486

Peattie, K., & Crane, A. (2005). Green marketing: Legend, myth, farce or prophesy? *Qualitative Market Research*, *8*(4), 357–370. DOI: 10.1108/13522750510619733

Popovich, N. (2024). Where E.V.s are (and aren't) taking off across the U.S. *The New York Times*. https://www.nytimes.com/interactive/2024/03/06/climate/hybrid-electric-vehicle-popular.html

Porter, M. E., & van der Linde, C. (1995, September-October). Green and Competitive: Ending the Stalemate. *Harvard Business Review*, 119–134.

PRNewswire. (2023). S&P Global Mobility Survey Finds EV Affordability tops Charging and Range Concerns in Slowing EV Demand. *S&P Global Mobility*. https://press.spglobal.com/2023-11-08-S-P-Global-Mobility-Survey-Finds-EV-Affordability-tops-Charging-and-Range-Concerns-in-Slowing-EV-Demand

Prothero, A. (1996). Environmental decision making: Research issues in the cosmetics and toiletries industry. *Marketing Intelligence & Planning, 14*(2), 19–25. DOI: 10.1108/02634509610110769

Rogers, E. M. (1962). *Diffusion of Innovations*. Free Press of Glencoe.

Schultz, P. W. (2022). Secret Agents of Influence: Leveraging Social Norms for Good. *Current Directions in Psychological Science, 31*(5), 443–450. DOI: 10.1177/09637214221109572

Schwartz, S. H. (1977). Normative influences on altruism. *Advances in Experimental Social Psychology, 10*, 221–279. DOI: 10.1016/S0065-2601(08)60358-5

Schwing, M., Reit, V., Selinka, S., Kuhn, M., & Meyer-Waarden, L. (2025). How Consumer Experience Shapes the Adoption of Autonomous Technologies: A Longitudinal Study With Shared Autonomous Vehicles. *Journal of Consumer Behaviour, 24*(3), 1133–1153. DOI: 10.1002/cb.2471

Seikh, M. R., & Chatterjee, P. (2025). Sustainable Strategies for electric Vehicle adoption: A confidence level-based interval-valued spherical fuzzy MEREC-VIKOR approach. *Information Sciences, 121814*. Advance online publication. DOI: 10.1016/j.ins.2024.121814

Shaheen, S. A., & Pan, A. (2024). Behavioral and sociodemographic impacts of carsharing. In *Handbook of Travel Behaviour* (pp. 339–363). Edward Elgar Publishing. DOI: 10.4337/9781839105746.00026

Statista. (2023). Global supply of EV batteries for recycling 2020-2040 [Dataset]. https://www-statista-com.eu1.proxy.openathens.net/statistics/1415407/ev-battery-recycling-supply-worldwide/

Steg, L., & Vlek, C. (2009). Encouraging pro-environmental behaviour: An integrative review and research agenda. *Journal of Environmental Psychology, 29*(3), 309–317. DOI: 10.1016/j.jenvp.2008.10.004

Stern, P. C. (1999). Information, incentives, and pro environmental consumer behavior. *Journal of Consumer Policy, 22*(4), 461–478. DOI: 10.1023/A:1006211709570

Summers, C. A., Smith, R. W., & Reczek, R. W. (2016). An audience of one: Behaviorally targeted ads as implied social labels. *The Journal of Consumer Research, 43*(1), 156–178. DOI: 10.1093/jcr/ucw012

Tanner, C. (1999). Constraints on environmental behaviour. *Journal of Environmental Psychology, 19*(2), 145–157. DOI: 10.1006/jevp.1999.0121

Tansini, A., Pavlovic, J., & Fontaras, G. (2022). Quantifying the real-world CO2 emissions and energy consumption of modern plug-in hybrid vehicles. *Journal of Cleaner Production, 362*, 132191. DOI: 10.1016/j.jclepro.2022.132191

Thøgersen, J., & Ölander, F. (2003). Spillover of environment-friendly consumer behaviour. *Journal of Environmental Psychology, 23*(3), 225–236. DOI: 10.1016/S0272-4944(03)00018-5

Uddin, S. M. F., Sabir, L. B., Kirmani, M. D., Kautish, P., Roubaud, D., & Grebinevych, O. (2024). Driving change: Understanding consumers' reasons influencing electric vehicle adoption from the lens of behavioural reasoning theory. *Journal of Environmental Management, 369*, 122277. DOI: 10.1016/j.jenvman.2024.122277 PMID: 39222587

UNEP. (2024). *United Nations Environment Programme Emissions Gap Report 2024*. https://www.unep.org/resources/emissions-gap-report-2024

Van Vliet, O. P., Kruithof, T., Turkenburg, W. C., & Faaij, A. P. (2010). Techno-economic comparison of series hybrid, plug-in hybrid, fuel cell and regular cars. *Journal of Power Sources, 195*(19), 6570–6585. DOI: 10.1016/j.jpowsour.2010.04.077

Verplanken, B., & Aarts, H. (1999). Habit, Attitude, and Planned Behaviour: Is Habit an Empty Construct or an Interesting Case of Goal-directed Automaticity? *European Review of Social Psychology, 10*(1), 101–134. DOI: 10.1080/14792779943000035

Verplanken, B., Aarts, H., Van Knippenberg, A., & Van Knippenberg, C. (1994). Attitude versus General Habit: Antecedents of Travel Mode choice. *Journal of Applied Social Psychology, 24*(4), 285–300. DOI: 10.1111/j.1559-1816.1994.tb00583.x

Verplanken, B., & Orbell, S. (2003). Reflections on Past Behavior: A Self-Report Index of Habit strength. *Journal of Applied Social Psychology, 33*(6), 1313–1330. DOI: 10.1111/j.1559-1816.2003.tb01951.x

Westaby, J. D. (2005). Behavioral reasoning theory: Identifying new linkages underlying intentions and behavior. *Organizational Behavior and Human Decision Processes, 98*(2), 97–120. DOI: 10.1016/j.obhdp.2005.07.003

World Bank Open Data. (2018). [Dataset]. https://datos.bancomundial.org/indicator/SP.URB.TOTL.IN.ZS?locations=ES

Young, W., Hwang, K., McDonald, S., & Oates, C. J. (2009). Sustainable consumption: Green consumer behaviour when purchasing products. *Sustainable Development (Bradford), 18*(1), 20–31. DOI: 10.1002/sd.394

Chapter 5
Sustainable AI Solution for the Transport System

Kashvi Chaturvedi
https://orcid.org/0009-0005-3192-3400
Ajeenkya D.Y. Patil University, India

Susanta Das
https://orcid.org/0000-0002-9314-3988
Ajeenkya D.Y. Patil University, India

ABSTRACT

Congestion, pollutants, and infrastructure stress are some of the growing issues faced by the transportation industry. This chapter analyzes present uses, difficulties, and future directions to examine sustainable AI solutions as a transformational force. It highlights carbon reduction and ethical governance while examining AI's involvement in autonomous vehicles, traffic optimization, predictive maintenance, and upcoming systems like the Hyperloop through a thorough literature analysis. From 1995 to 2025, bibliometric studies show changing research patterns that emphasize multidisciplinary connections with Industry 4.0, healthcare, and climate action. Key findings highlight AI's ability to address obstacles including data quality, cybersecurity, and equity while aligning with UNSDGs for resilient, low-emission mobility. In order to promote smarter, greener transportation ecosystems, this chapter advocates for the ethical and scalable deployment of AI by filling gaps in systemic integration.

DOI: 10.4018/979-8-3373-4495-9.ch005

1. INTRODUCTION

A vital pillar of modern civilization, the global transportation industry is deeply interwoven with social mobility, economic expansion, and global connectedness. It is the backbone of business, allowing the smooth transportation of commodities across long distances and supporting complex supply chains that are necessary for both domestic and foreign trade (Garg & Sharma, 2022; Zhang et al., 2020). In addition to its economic role, transportation plays a vital role in facilitating social unity by connecting communities, granting access to necessary services, and creating chances for both personal and professional growth. A robust and effective transportation infrastructure is essential for development and prosperity, from everyday commutes within cities to the intricate logistics of global trade (Abduljabbar et al., 2019). The sector's fundamental significance emphasizes the need for ongoing innovation to guarantee its long-term sustainability and to solve the urgent issues it is currently facing.

Notwithstanding its vital role, the transportation system is under tremendous and increasing strain due to a complicated web of interrelated problems. Rapid urbanization has resulted in a notable rise in vehicle density and, as a result, inescapable traffic congestion (Al-Hussaini et al., 2021; Gao et al., 2020). In addition to causing significant financial losses through missed output and postponed shipments, this congestion worsens environmental problems by increasing fuel and idle time. Additionally, transportation continues to be a key contributor to air pollution and a leading source of global carbon emissions, which has a direct impact on public health and accelerates negative climate change (Guo et al., 2021). Both drivers and pedestrians are at serious danger for injury due to the alarmingly high frequency of incidents caused by the sheer volume of vehicle traffic. At the same time, existing infrastructure—such as public transportation networks, roads, and bridges—is frequently aging and finding it difficult to meet demand, requiring major capital investment and ongoing maintenance (Mohanty et al., 2022; Li et al., 2020). These systemic problems show how urgently a paradigm change is required, departing from traditional methods that have not been able to keep up with the needs of a world that is changing quickly.

Given this, artificial intelligence (AI) has become a deeply revolutionary technology that provides a strong and adaptable toolkit to support the sustainability objectives of transportation systems and address systemic issues (Wang et al., 2018; Bhatnagar & Sastry, 2020). Intelligent decision-making that vastly outperforms conventional, rule-based systems is made possible by AI's ability to process and analyze enormous, complicated datasets in real-time. By eliminating vehicle pauses and idle time, AI-driven traffic management systems, for example, can optimize traffic signal timing and flow in real-time, greatly lowering congestion and fuel consumption (Chen et al., 2020; Dhivya & Gowthami, 2020; Liu et al., 2021). Similar to this, AI-based vehicle routing algorithms can identify the most effective routes to cut down on trip distances and carbon emissions in the logistics and freight delivery industry. This improves fleet management and lowers operating costs (Luo & Zhang, 2019; Ghasemi & Zarrabi, 2022; Sarkar & Pal, 2020). AI is essential for improving safety and dependability in addition to operational efficiency. By using machine learning to predict equipment failures in automobiles and rail systems, predictive maintenance models, for instance, enable preemptive repairs that avert mishaps and expensive service interruptions (Zou et al., 2021; Li et al., 2020; Shah et al., 2021). The introduction of advanced AI-powered autonomous cars has the potential to completely transform transportation by lowering accident rates, enhancing traffic flow, and facilitating the development of more connected and energy-efficient transportation networks (Vasan & Balamurugan, 2021; Goudarzi & Mohammadi, 2021). AI is being used by even new technologies, like as the Hyperloop, to optimize their operational models for energy efficiency, speed, and safety (Hofstetter et al., 2019). The development of smarter, safer, and more sustainable transportation ecosystems is largely dependent on AI-driven solutions (Wang et al., 2021).

Significant research gaps and problems exist in the subject of "sustainable AI solutions for transport systems" despite the apparent potential of AI. The promise of individual AI applications has been shown in numerous studies, but thorough studies evaluating the long-term, systemic effects of integrating multiple AI solutions across a whole transportation network are conspicuously lacking (Guo et al., 2021). More research is still needed to fully understand these models' scalability, especially in diverse and dynamic metropolitan environments, and the necessity of ensuring equity and fairness in their application. The development of sustainable AI also entails tackling the energy usage of the AI systems itself and making sure that they are deployed ethically, in addition to improving operational efficiency.

Therefore, the goal of this chapter is to present a thorough and compiled summary of the state of AI applications in transportation today. By combining information from other fields, showcasing effective applications, and outlining important obstacles and future prospects, this work seeks to close the current research gaps. This chapter aims to serve as a foundational text for researchers, policymakers, and industry practitioners by critically analyzing how AI can be used to support fundamental sustainability objectives. This will help guide the development of AI solutions that are not only technologically sophisticated but also socially and environmentally responsible. In the end, this chapter also presents bibliometric analysis on Sustainable Transportation depicting the trend of research in the field.

The world transport systems have never experienced the challenges that are being experienced today in rapid urbanization, population increase as well as increased mobility when compared to the past (Quiroga, 2012; Shladover, 2017). Traffic jams, greenhouse gas emissions, road safety hazards, and infrastructure pressure have been longstanding issues that have rendered the conventional models of transport unsustainable (Shladover, 2017). Such issues have developed a pressing need of smart, adaptable, and greener mobility solutions.

The concept of Artificial Intelligence (AI) has become the revolutionary technology that can be used in overcoming most of these structural constraints through the use of data-driven decision making, real-time optimization of systems, and proactive analytics (Floridi et al., 2018; Saadi et al., 2023). AI will become the basis of creating more efficient, safer, and sustainable transportation networks, allowing multisided resources distribution, smart traffic regulation, and automated piloting.

Nevertheless, the implementation of AI-based transportation is extremely uneven on the distribution of regions worldwide. Advanced digital infrastructure and data ecosystems are enjoyed in developed countries and developing countries still struggle to overcome technological, financial, and institutional restrictions (OECD, 2023). This gap endangers the growth of inequality in access to sustainable mobility and emphasizes the need to make technologies significantly inclusive and collaborate on a global level (UNCTAD, 2023; Saadi et al., 2023).

Furthermore, the use of AI in transportation has to be coordinated with the international policies, including the United Nations Sustainable Development Goals (SDGs) and the Paris Climate Agreement (United Nations, 2015; UNFCCC, 2016). Mobility with AI will be a part of the sustainable infrastructure, low-carbon city infrastructure, and mitigation approaches to climate change (IEA, 2024).

2. LITERATURE REVIEW

2.1 AI Applications in Transportation

A paradigm change from conventional, reactive management to a proactive, data-driven operational style is represented by the introduction of artificial intelligence (AI) into the transportation industry. The creation of intelligent systems that improve productivity, security, and sustainability has been made possible by AI's ability to handle and analyze enormous, complex datasets from a variety of sources, including as sensors, cameras, and vehicle-to-everything (V2X) communication networks (Wang et al., 2018). This part offers a thorough analysis of the primary AI applications in operational automation, traffic management, predictive maintenance, and safety procedures, emphasizing how they contribute to the development of a more sustainable transportation ecosystem.

Optimizing traffic flow is one of the most direct and significant uses of AI in transportation, directly addressing the pressing issues of pollution and congestion.

According to Abduljabbar et al. (2019), traditional traffic management frequently fails to adjust to dynamic changes in traffic volume and density because it depends on fixed-time signals or basic responsive logic. On the other hand, AI-driven systems use real-time data to apply adaptive and intelligent control mechanisms. In order to minimize vehicle delays and reduce idling, Intelligent Traffic Light Systems (ITLS) employ deep reinforcement learning to assess real-time traffic patterns and dynamically modify signal timings (Chen et al., 2020; Liu et al., 2021). In addition to improving traffic flow, this adaptive management greatly lowers fuel usage and the resulting carbon footprint. An AI-driven traffic management system for smart cities may result in better traffic flow and a more economical use of road infrastructure, according to research by Dhivya and Gowthami (2020).

AI models are being employed more and more for congestion prediction and reduction in addition to active control. With the ability to recognize intricate non-linear patterns in huge datasets, deep learning models are highly accurate at predicting traffic congestion. Gao et al. (2020), for example, created a deep learning-based model to forecast urban traffic congestion, offering useful information for preventative traffic control. Similar to this, Mishra and Garg (2022) suggested a deep learning method for predicting traffic flow, which can help traffic authorities and city planners foresee and alleviate bottlenecks before they arise. In order to create a more robust and sustainable urban transportation system, these predictive capabilities enable dynamic road pricing, traffic rerouting via less crowded routes, or the activation of particular public transportation lines to absorb demand. Additionally, fleet management and individual drivers can optimize their trips with AI-enabled real-time navigation systems and dynamic route planning. By providing the most efficient routes based on real-time traffic, weather, and road conditions, these systems cut down on fuel use and trip time. AI-driven dynamic pricing for ride-sharing services can encourage travel at off-peak times, distributing demand and reducing traffic even further (Teng & Guo, 2020). These AI applications work together to create a road network that is more responsive and effective, which lowers emissions and promotes a more sustainable use of energy resources.

Vehicle and infrastructure maintenance is a resource-intensive, frequently reactive procedure that raises operating expenses and system outages. A paradigm changes from a time-based or breakdown-based strategy to a data-driven, condition-based one is represented by AI-based predictive maintenance. In order to anticipate possible equipment problems before they happen, machine learning algorithms are used to evaluate data from sensors, onboard diagnostics, and previous maintenance records. For instance, Li et al. (2020) showed how effective an AI-based predictive maintenance system was for urban rail transit, enabling prompt repairs and lowering the possibility of service interruptions. Similar to this, Zou et al. (2021) offered a machine learning-based method to forecast battery and component failures for electric vehicles, which have unique maintenance requirements. This ensures vehicle dependability and increases asset longevity. By doing maintenance only when absolutely necessary, this proactive strategy not only improves the safety and dependability of public transportation systems but also reduces waste and maximizes resource utilization. Additionally, Shah et al. (2021) emphasized how predictive maintenance for autonomous cars might improve safety and dependability, which is essential for the broad use of these technologies.

Automation powered by AI is radically changing how the transportation industry operates, increasing productivity and lowering environmental impact. The fields of scheduling and fleet management have seen some of the biggest developments. By taking into account several factors at once, such as distance, delivery time, fuel efficiency, and vehicle capacity, AI algorithms can improve complicated logistics challenges including last-mile delivery and freight truck routing (Luo & Zhang, 2019). As a result, carbon emissions and needless travel are significantly decreased. Sarkar and Pal (2020), for instance, investigated how fleet management powered by AI can successfully lower the carbon impact of last-mile delivery. Artificial intelligence (AI) algorithms can be used in the public transportation sector to optimize bus and rail timetables in response to real-time passenger demand, which would improve service delivery and lower operating costs. According to Wang et al. (2021) and Garg and Sharma (2022), smart public transport management systems use artificial intelligence (AI) to manage fare collection, give passengers real-time information, and coordinate various modes of transportation, all of which encourage the use of shared and public mobility options rather than private vehicles. In addition to increasing transportation efficiency, automation of these procedures promotes the transition to more scalable and sustainable mobility solutions.

Any transportation system must prioritize safety, and artificial intelligence is becoming more and more important in improving emergency response and accident prevention. AI-powered systems track road conditions and driver behavior in real time using computer vision, sensor data, and behavioral analysis. These technologies can identify indicators of driver fatigue or distraction, send out alerts in a timely manner, and in more sophisticated autonomous cars, even take over to stop an impending collision (Vasan & Balamurugan, 2021). As previously mentioned, (Shah et al., 2021), the application of AI in predictive maintenance also improves safety by averting vehicle malfunctions that can cause collisions. Additionally, by evaluating traffic data to identify the quickest routes for emergency vehicles, AI can optimize emergency response and guarantee a faster reaction time. By combining artificial intelligence (AI) with vehicle-to-everything (V2X) communication, cars may exchange data on traffic patterns, risks, and road conditions, forming a collective intelligence that improves user safety on the road. Thus, AI-enabled solutions are not only improving transportation efficiency but also fostering a more secure and safe environment for all.

2.2 General Overview

The available literature evidences that the use of AI in various areas of transportation is rapidly developing, among which are traffic management, autonomous driving, predictive maintenance, and smart logistics (Shladover, 2017). Several articles claim the solution to the efficiency of operations and safety performance of reinforcement learning, machine learning, and deep learning models (Quiroga, 2012). These researches have great technical support to the idea that AI can make a great contribution to improving the functionality of the transportation systems.

Having said that, even with the scope of the body, the current research is largely descriptive, as it talks little about system sustainability over time (Floridi et al., 2018). The inconsistencies in the datasets, experimental conditions, and evaluation methods introduce differences in the reported results. More so, little focus is placed on governance systems, ethical application, and applicability to a variety of social economic conditions (OECD, 2023). Such gaps indicate that more critical and comparative research approaches are required in this sphere.

2.3 AI for Road Safety and Autonomous Vehicles

The development of artificial intelligence (AI) has made it a key player in the effort to create safer transit systems. Road accidents are mostly caused by human mistake, but AI-driven technologies provide a revolutionary remedy by boosting vehicle perception, facilitating proactive safety measures, and improving decision-making. The basic role of artificial intelligence (AI) in autonomous vehicles (AVs) is examined in this part, along with the technologies that allow AVs to negotiate challenging situations and the wider effects on road safety, policy, and public trust. The development of completely autonomous systems represents a fundamental change in the definition of mobility itself, not just a technical breakthrough (Wang et al., 2018).

The Society of Automotive Engineers (SAE) often divides the development of autonomous vehicles into six stages, ranging from Level 0 (no automation) to Level 5 (complete automation). The growing complexity of AI is the primary force behind this trip. AI is used in the early stages, like Level 1 and 2, for particular driver-assistance functions like adaptive cruise control and lane-keeping support. The smooth integration of a range of AI technologies is necessary for the advancement to higher levels, when the AI assumes increasingly difficult driving tasks (Vasan & Balamurugan, 2021). The pinnacle of this AI-driven development is Level 5 autonomy, which promises to revolutionize freight transportation and urban mobility by enabling a vehicle to function in any situation without the need for human involvement.

Computer vision and deep learning are essential to an autonomous vehicle's capacity to "see" and comprehend its surroundings. These technologies use camera data to track, identify, and categorize things in real time. To identify people, bicycles, other vehicles, and stationary objects with exceptional precision, deep neural networks—in particular, convolutional neural networks, or CNNs—are trained on enormous image datasets (Mishra & Garg, 2022). Navigating dynamic and unpredictable urban surroundings requires this visual perception. In a similar vein, computer vision systems are necessary for the recognition and interpretation of road markings, traffic signs, and traffic lights. This enables the AI to comply with traffic regulations and react correctly to signals that change. Since a single misidentification might have disastrous results, the dependability of these systems is crucial, making ongoing model validation and improvement an important field of study.

Despite being a potent information source, computer vision is vulnerable to issues like dim lighting, bad weather, and visual obstructions. Sensor fusion, a method that integrates information from several redundant sensors to produce a more comprehensive and accurate model of the vehicle's environment, is used by autonomous systems to get around these issues. In addition to cameras, this also incorporates inputs from Inertial Measurement Units (IMUs), RADAR (Radio Detection and Ranging), and LIDAR (Light Detection and Ranging). While RADAR is excellent at determining an object's distance and speed in any weather, LIDAR offers a very accurate three-dimensional image of the surroundings. A strong and thorough view of the surroundings is produced by combining data from these various sensors, frequently enhanced by GPS and information from Internet of Things devices (Garg & Sharma, 2022; Wang et al., 2018). The system can make safer and better decisions by fusing these data streams using artificial intelligence. Additionally, a vehicle's awareness is further enhanced by the idea of smart infrastructure, which uses AI and IoT to provide real-time data on traffic flow, parking availability, and road conditions (Al-Hussaini et al., 2021). This results in a more intelligent and cohesive transportation network (Mohanty et al., 2022).

What really distinguishes an autonomous vehicle's intelligence is its capacity to make safety-critical judgments in real time. The AI needs to process the information and carry out a logical activity after observing its surroundings. This is accomplished by using complex decision-making frameworks, which frequently make use of behavioral cloning and reinforcement learning. AI agents are trained using reinforcement learning to maximize a "reward" function, which in the context of driving might be a mix of energy efficiency, safety, and speed (Teng & Guo, 2020). The AI learns to handle challenging circumstances and respond to unanticipated events by modeling millions of scenarios. In contrast, behavioral cloning enables an AI model to replicate human-like driving behavior by training it on data from human drivers. These frameworks are essential for moving from perception to action, enabling the car to stop, change lanes, and plan routes in a predictable and safe way (Bhatnagar & Sastry, 2020).

AI's capacity for decision-making also presents difficult moral conundrums with no simple solutions. The "trolley problem," in which an AV has to make a potentially harmful judgment, is an often-discussed scenario. Even though these situations are uncommon in real life, they emphasize how important it is for autonomous systems to have a clear ethical framework. Whether a car should prioritize a pedestrian, another vehicle, or a passenger in a no-win situation depends on social values and needs to be carefully managed. Scholars and policymakers are striving to create transparent algorithms and decision hierarchies that conform to legal and social norms. However, a major obstacle to the broad, moral application of Level 5 autonomy is still the lack of uniform international legislation on the subject.

The capacity of a vehicle to interact with its environment through Vehicle-to-Everything (V2X) connectivity significantly improves AI-driven safety beyond its onboard sensors. Vehicle-to-Pedestrian (V2P), Vehicle-to-Infrastructure (V2I), and Vehicle-to-Vehicle (V2V) networks are included in this. Artificial intelligence (AI) systems can use this data to predict what will happen outside of a vehicle's field of vision. The AI can anticipate and respond faster than a human driver, for instance, when a car receives a signal from a traffic light about an approaching red light or a warning from another vehicle about an abrupt brake. A safer and better-coordinated traffic flow is produced by this collective intelligence, in which infrastructure and automobiles exchange data in real-time, greatly lowering the chance of collisions (Goudarzi & Mohammadi, 2021). This trend is anticipated to pick up speed with the rollout of 5G networks, which will offer the high bandwidth and low latency required for such real-time data transmission.

AI has significant potential effects on road safety. Human error, such as being distracted, tired, or drunk, is the cause of the great majority of road accidents. Road accidents and fatalities can drastically decline as a result of AI-powered solutions that can eliminate or drastically minimize these risks (Vasan & Balamurugan, 2021). Liability models for accidents will change as AI replaces more driving tasks, shifting the blame from the human driver to the software developers and automakers. The insurance sector will need to review its risk models as a result of this, which has important ramifications. To ensure that the technology is used responsibly and for the public's benefit, regulatory organizations from all over the world are trying to create new frameworks and standards to verify the safety and dependability of autonomous vehicles (Wang et al., 2021).

The AI-based traffic management systems make use of real-time sensor information, computer vision, and predictive modeling to complete traffic signal timing optimization, traffic congestion control, and general traffic flow(Quiroga, 2012). Due to intelligent transportation systems, the dynamism of the system to react to the changing traffic conditions will reduce the amount of idling, fuel consumption, and vehicular emissions (Shladover, 2017). These possibilities render AI an influential means of improving mobility in the city.

Nevertheless, most of the current implementations only run pilot stages or in-simulations and how they will perform in the complicated, real-life setting remains under-researched (Saadi et al., 2023). Problems of interoperability of the systems, their maintenance in the long term, and equity in traffic prioritization are seldom discussed. To create scalable, resilient and context-aware AI systems, which can be effectively deployed in varied urban and regional contexts is an issue that should be addressed in future studies (Floridi et al., 2018).

2.4 AI in Emerging Transport Systems

Innovation in transportation is no longer limited to the conventional areas of rail and road. Emerging transport systems are positioned to be vital to the future of connectivity as the demand for quicker, more sustainable, and more effective movement increases globally. High-speed automation and energy efficiency are the cornerstones of these next-generation systems, like the Hyperloop, and artificial intelligence's potential is essential to their feasibility. Using the Hyperloop as a primary case study, this section explores the revolutionary role of AI in these emerging systems, demonstrating how essential AI is to their sustainability, operation, and design.

With its goal of transporting people and commodities through low-pressure tubes at nearly supersonic speeds, the hyperloop technology is a radical break from traditional modes of transportation (Hofstetter et al., 2019). This method has the potential to be substantially more energy-efficient than high-speed rail or aviation by greatly lowering air resistance and friction. However, a degree of accuracy, control, and real-time optimization that is beyond human capacity is required for the successful deployment of such a system. At this point, AI stops being a supplemental technology and starts to function as a foundational one. A sophisticated, self-regulating system is necessary because to the intricacy of controlling pods, preserving a near-vacuum atmosphere, and guaranteeing passenger safety at speeds exceeding 1000 km/h. For a large-scale Hyperloop network to be efficient and sustainable, powerful AI must be used in its design, construction, and operation (Goudarzi & Mohammadi, 2021).

For example, AI is ideally adapted to address extremely difficult optimization problems, like as the design and route planning for an Hyperloop network. To choose the best route for the tube network, the AI must concurrently take into account a wide range of factors, such as land use, passenger demand, energy consumption, and geographic restrictions. This is a dynamic issue that necessitates ongoing reassessment in order to adjust to shifting circumstances and operational objectives (Guo et al., 2021). In a similar vein, AI is essential to the system's energy optimization. The power needed to keep the vacuum inside the tubes constant and to accelerate and decelerate pods is enormous. By evaluating operational data and precisely modifying the braking and propulsion systems, AI algorithms can control this energy use in real-time. AI can, for instance, employ regenerative braking to recover a sizable amount of the kinetic energy and schedule pod departures to reduce peak energy demands (Hofstetter et al., 2019). This strategy directly supports the objectives of climate action by lowering the system's overall carbon footprint and operating expenses.

The high-speed environment and the vital requirement to preserve the structural integrity of the vacuum tubes make maintenance more difficult for Hyperloop systems. Predictive maintenance models powered by artificial intelligence, as those for urban rail transit (Li et al., 2020), are essential. Real-time data on temperature, pressure, vibrations, and structural stress is gathered by sensors mounted on the pods and imbedded throughout the tube network. Artificial intelligence (AI) algorithms can examine this data to find irregularities and anticipate possible tube or pod failures before they become dangerous. This proactive maintenance strategy reduces downtime, avoids catastrophic failures, and guarantees the system's long-term dependability and safety (Goudarzi & Mohammadi, 2021). AI is also essential for operational effectiveness, especially when it comes to the challenging process of scheduling and overseeing a sizable fleet of pods. In order to optimize throughput, reduce delays, and guarantee a smooth travel experience for passengers, these algorithms can schedule the departure and arrival of pods. Managing the network's enormous traffic volume and speed requires this degree of automation. AI systems are essential for passenger safety and security in addition to operational efficiency. With a speed and accuracy that human operators cannot match, they are able to monitor pod environs, identify probable dangers, and control emergency response procedures.

There is a lot of research being done on the possible scalability and environmental impact of new transportation systems like the Hyperloop. An Hyperloop system would be a very sustainable option for intercity transport due to its low energy consumption and long-term operational efficiency, even though it demands a large initial investment (Hofstetter et al., 2019). The Hyperloop could drastically cut the transportation sector's carbon emissions by providing a fast, carbon-neutral substitute for long-distance driving and short-haul aircraft. The secret to releasing this potential and bringing these futuristic ideas from the drawing board to life is the effective integration of AI as a fundamental part of its functioning and design.

Self-driving and connected cars are one of the most radical applications of AI in the transportation industry. These systems have proven that with impeccable sensor fusion, machine learning, and real-time decision-making algorithms, a substantial decrease in human error and increased road safety can be achieved (Shladover, 2017; Bonnefon et al., 2016). AI can be used to make vehicles recognize their surroundings and anticipate possible risks and proactively drive (Saadi et al., 2023).

Although technical progress has been made, major issues surrounding the subject are connected with the trust of the population, legal accountability and moral judgment in the life-threatening situations (Bonnefon et al., 2016; Floridi et al., 2018). According to the current literature, the issue of uneven societal acceptance and regulatory preparedness is observed in different countries (OECD, 2023). These issues have suggested that effective implementation of autonomous vehicles will not be based on technical reliability alone, but also effective governance structures and societal legitimacy.

2.5 Sustainability and Ethical Considerations

The discussion of AI's application in transportation goes far beyond operational effectiveness and technical viability. Its effect on sustainability and the intricate ethical issues raised by the application of intelligent systems are important aspects of this conversation. Examining AI's potential to promote a more ecologically friendly transportation ecology while also making sure that its implementation is fair, transparent, and just is crucial as the technology becomes more widely used. This section summarizes the academic viewpoints on these two issues, examining how AI might be used to lessen transportation's environmental impact and how the sector is dealing with basic moral conundrums.

AI provides a potent arsenal for reducing the transportation sector's substantial environmental effect, especially its carbon emissions. AI-powered energy-efficient mobility is one of the most obvious uses. AI can reduce the amount of energy used by automobiles and public transportation systems by utilizing optimization algorithms and predictive analytics. For instance, AI models may optimize train speeds and braking in urban rail transportation by using reinforcement learning, which can save a substantial amount of energy (Huang et al., 2019). Additionally, AI-driven navigation systems can suggest eco-routing—routes that reduce fuel consumption by avoiding steep inclines, heavy traffic, and excessive acceleration and deceleration—in the context of road transportation. AI may actively guide the transportation system toward greater environmental sustainability by making a vehicle's energy use a key optimization parameter.

Another way AI supports sustainability is through promoting multimodal transportation systems and shared mobility. In ride-sharing services, AI platforms can effectively match drivers with vehicles, lowering the number of automobiles on the road and alleviating traffic in cities (Teng & Guo, 2020). Beyond ride-sharing, AI makes it easier to integrate many forms of transportation, including buses, trains, bicycles, and walking, by giving real-time schedule and availability information. This significantly lowers per-capita emissions by encouraging people to choose shared or public transportation over private automobiles. By optimizing these multimodal networks, AI-enabled solutions can promote sustainable transportation in smart cities, according to a thorough assessment by Wang et al. (2021). Additionally, by identifying the most fuel-efficient routes and coordinating the usage of electric vehicles, AI algorithms in the logistics domain can optimize urban freight transportation, directly reducing carbon emissions in the supply chain (Ghasemi & Zarrabi, 2022; Zhang et al., 2020). In line with the global climate action goals (SDG 13), the combined effect of these applications is a significant reshaping of mobility patterns towards more sustainable solutions.

Despite being essential to their operation, AI systems' massive data collecting and processing create serious ethical issues, such as privacy, responsibility, and transparency. AI-powered transportation systems gather private information about people's whereabouts, travel patterns, and, occasionally, biometric data. Because breaches could have serious repercussions for privacy and personal safety, it is crucial that sensitive data be managed securely. Strong data governance frameworks that guarantee anonymity, offer unambiguous consent procedures, and specify the intended use of data are desperately needed. Furthermore, a significant obstacle to accountability is the "black box" character of many AI systems, where the decision-making process is unknown even to its developers. Without a clear grasp of the algorithm's reasoning, it might be challenging to assign blame when an AI-driven system makes a mistake, such as when an autonomous car causes an accident.

The possibility of prejudice in AI systems, which can disproportionately affect particular communities, is another important ethical consideration. Transport access and social fairness may be impacted, for instance, if a traffic optimization algorithm unintentionally prioritizes traffic flow in wealthy districts while ignoring infrastructure in low-income ones. This may occur if the AI models' training data is not representative of all demographic groups. To tackle this, a concentrated effort is needed to guarantee data diversity and expose algorithms to stringent bias-detection audits. The significance of developing inclusive, AI-driven smart infrastructure that benefits every member of the community is emphasized by Mohanty et al. (2022). In order to prevent AI-driven transportation from widening already-existing societal divides, it is necessary to look beyond a merely technological perspective and take into account the social and human ramifications of these systems.

AI-driven transportation solutions must be designed and implemented with community involvement and inclusive design in order to promote public trust and guarantee fair results. One of the biggest obstacles to the broad use of AI technologies, especially autonomous vehicles, is public acceptability. Implementation might be significantly hampered by resistance to change and a general sense of safety dangers. Policymakers and developers may address public concerns, foster trust, and create solutions that truly serve the needs of various populations by involving communities early in the development process. AI-enabled systems should be designed collaboratively, taking into account input from a variety of stakeholders, such as the public, regulators, and urban planners. This guarantees that the systems are socially conscious, technically sound, and in line with the objectives of building inclusive and sustainable cities (Wang et al., 2021). In order to guarantee that ethical issues are not an afterthought but rather an essential component of the development lifecycle, it is imperative that effective governance frameworks be established (Bhatnagar & Sastry, 2020).

The increasing application of AI to transportation presents a major ethical and governance concern regarding the privacy of data, the transparency of the algorithm, accountability, and bias (Floridi et al., 2018). The AI systems are becoming more and more involved in the safety-related decisions, and it is a crucial point to make sure that their work is explainable and in accordance with the values of humans (Bonnefon et al., 2016).

International AI standards and regulatory measures are among global governance projects that seek to provide the framework to responsible AI application (OECD, 2023). But the interoperability and global scale is problematic as harmonization is not achieved across countries. Enhancing global cooperation and integrating both ethics and system design make it necessary to create the population trust and implement the methods sustainably (Floridi et al., 2018).

2.6 Challenges and Limitations

Even while AI has enormous potential to revolutionize sustainable transportation, there are still several obstacles in the way of its successful and broad implementation. These restrictions cover a wide range of topics, including complicated socioeconomic issues, legal and regulatory inadequacies, infrastructure limitations, and technical difficulties. To guide research, policy-making, and implementation methods that will open the door to a really integrated and sustainable transport ecosystem, a critical knowledge of these obstacles is required. This section explores the fundamental obstacles that need to be overcome in order to fully utilize AI in this field.

The resilience of the underlying technology and the availability and quality of data provide the technological basis of AI-driven transport systems. Ensuring data quality and platform integration is one of the biggest obstacles. AI models are heavily reliant on data for uses such as autonomous driving, predictive maintenance, and traffic management (Sarkar & Pal, 2020). However, the data frequently comes from a range of sources, such as legacy infrastructure, mobile devices, and Internet of Things sensors, which can lead to incompleteness, inconsistencies, and a lack of standards. It is a difficult technical undertaking that calls for substantial computational resources and sophisticated data engineering skills to integrate these different datasets into a cohesive, real-time system (Gao et al., 2020). Real-time processing of this massive amount of data is essential for AI-driven systems to function well, especially in dynamic metropolitan areas. Processing lag can make autonomous driving judgments or real-time traffic forecasts erroneous, which could be dangerous.

Furthermore, there are serious cybersecurity risks because AI-enabled infrastructure is networked. Malicious attacks on the software of an autonomous car or traffic control system might cause catastrophic failures, safety risks, and extensive disruption. From in-car systems to cloud-based platforms that control traffic flow and fleet management, the attack surface is extensive. Strong encryption protocols, sophisticated threat detection models, and ongoing monitoring are necessary to protect these systems (Guo et al., 2021; Wang et al., 2018). These systems' susceptibility is still a significant operational and technological problem that calls for constant research and funding, especially in an era of increasingly complex cyberthreats.

Significant economic and infrastructure constraints frequently impede the adoption of AI-enabled solutions. One of the biggest issues is the high implementation and maintenance costs. It takes a significant financial outlay to install the required hardware, such as road sensors, smart traffic signals, and electric vehicle charging infrastructure. For instance, installing a dense network of cameras, sensors, and communication devices is necessary for a full AI-driven traffic control system, but many municipalities may find this to be unaffordable (Dhivya & Gowthami, 2020). Long-term financial strain is also posed by the expense of maintaining and improving this infrastructure, as well as the continuous operational expenditures of data processing and algorithm upkeep. This cost barrier adds to the restricted implementation in poor nations, where adopting advanced AI transport solutions is challenging due to a lack of existing digital infrastructure and financial constraints. This exacerbates a worldwide digital divide in transportation innovation since the advantages of these sustainable technologies are frequently concentrated in a small number of developed urban centers (Garg & Sharma, 2022).

The broad use of AI in transportation may be hampered by unclear and inconsistent governance, even with the infrastructure and technology in place. The absence of common AI governance policies is a significant obstacle. The speed at which technology is developing frequently surpasses authorities' capacity to provide thorough frameworks for the development and application of AI. Developers and operators may experience uncertainty as a result of fragmented and varied policies across many countries caused by this regulatory vacuum (Wang et al., 2021). It is challenging to scale solutions and gain public trust when there is no worldwide agreement on data protection, algorithmic transparency, and ethical AI use.

Furthermore, liability concerns with autonomous systems continue to be a major ethical and legal conundrum. Determining legal culpability in an accident involving an autonomous vehicle is difficult. Who is at fault—the passenger, the owner of the car, the software developer, or the manufacturer? These novel situations are beyond the capabilities of the current legal structures, which are predicated on human drivers. For autonomous vehicles to be widely commercialized and accepted by the public, these liability issues must be resolved (Vasan & Balamurugan, 2021).

The effective incorporation of AI into transportation systems depends on resolving significant societal issues in addition to technological and legal obstacles. Opposition to change and job relocation are two examples of such issues. Many human-centric jobs in the transportation industry could become obsolete due to automation of driving, logistics, and fleet management operations, which would cause job losses and economic distress. Political and popular opposition to the deployment of these technologies may result from this concern of job displacement.

Another significant obstacle is the public's sense of safety and trust in AI-driven solutions. Even while statistics show that autonomous cars can be safer than automobiles driven by people, high-profile collisions or system malfunctions frequently damage public confidence. Strong accountability procedures, transparent communication, and a convincing illustration of these systems' safety advantages are necessary to foster trust in them. Convincing the public that the technology is safe is just as difficult as making it safe (Mohanty et al., 2022).

2.7 Future Directions

As seen in the sections above, artificial intelligence (AI) has the potential to revolutionize the transportation industry by solving long-standing problems and opening the door to a more sustainable system. But the current state of AI use in transportation is just the start of a significant development. In order to create a robust, intelligent, and linked mobility ecosystem, future research and development must concentrate on an integrated, holistic strategy that transcends single-point solutions. The development of intelligent multimodal systems, the convergence of important technologies, and the alignment with global sustainability goals are highlighted in this part, which also defines the major future directions for AI in transportation.

AI-driven transportation's future depends on how well it integrates with other cutting-edge technologies, particularly 5G networks and the Internet of Things (IoT). A genuinely smart transportation ecosystem where people, infrastructure, and cars are all connected in real-time will be created by the confluence of AI and IoT. Large volumes of data will be continuously collected by IoT devices, such as sensors installed in public transportation vehicles, traffic lights, and roadways. This data will subsequently be processed by AI algorithms to produce dynamic, real-time insights that enhance sustainability and efficiency. For instance, IoT sensor data can be used by AI-based smart parking management systems to direct vehicles to open spaces, lowering emissions and traffic caused by searches (Al-Hussaini et al., 2021). Similar to this, AI and IoT can be used together to track the condition of vehicles and forecast maintenance requirements, especially for newer technologies like electric cars (Zou et al., 2021; Shah et al., 2021). By reducing vehicle down-time and extending the lifespan of transportation assets, this proactive approach to maintenance—made possible by real-time data from IoT sensors—contributes to increased sustainability.

Communication dependability and speed are critical to the effectiveness of this networked ecosystem. At this point, 5G's involvement becomes essential. 5G networks will be the foundation for the real-time data transfer needed by sophisticated AI systems because of their tremendous bandwidth and extremely low latency. For safety and efficiency, autonomous vehicles must be able to interact in milliseconds with pedestrians (V2P), traffic infrastructure (V2I), and each other (V2V). A new era of cooperative autonomous driving and traffic management will be made possible by 5G, enabling coordinated decisions that maximize traffic flow throughout a metropolis (Mishra & Garg, 2022). The entire potential of a smart city environment will be unlocked by the convergence of AI, IoT, and 5G, transforming sustainable transportation from an objective to a fundamental aspect of urban life.

Future studies must concentrate on creating intelligent multimodal systems rather than just maximizing specific forms of transportation. This entails the smooth fusion of air, sea, rail, and road transportation into a unified system. The core of this integration will be AI algorithms that use predictive analytics to optimize connections between various modes, reduce travel times, and improve the experience of passengers. An AI-powered platform might, for example, recommend a trip that combines a high-speed rail, a shared autonomous vehicle, and a last-mile electric scooter—all of which would be efficient and have a low carbon footprint.

Sophisticated multi-agent reinforcement learning models will be necessary for such systems, in which every form of transportation is viewed as an intelligent agent cooperating to accomplish a global goal, such lowering carbon emissions or overall traffic (Liu et al., 2021). With this strategy, a single, integrated mobility network replaces fragmented transportation systems. These systems will greatly aid in lowering traffic, enhancing accessibility, and building a more robust transportation infrastructure by strategically controlling the movement of people and products among various forms of transportation.

Numerous Sustainable Development Goals (SDGs) of the UN are closely associated with the idea of a sustainable AI-driven transportation system, offering a clear foundation for further study and policy. The creation of AI-powered transportation solutions is closely related to:

- Sustainable Cities and Communities is SDG 11. AI-powered solutions that optimize urban freight distribution, lessen traffic, and enhance public transportation help to create more inclusive, secure, resilient, and sustainable cities.
- SDG 13: Addressing Climate Change. AI can aid in the fight against climate change by enabling more effective logistics and encouraging low-carbon travel options, which will considerably reduce greenhouse gas emissions from the transportation sector (Luo & Zhang, 2019).
- SDG 9: Infrastructure, Industry, and Innovation. AI is a new area of innovation, and its use in transportation advances technology, encourages inclusive and sustainable industry, and helps create robust infrastructure.

The ultimate goal of AI in transportation is to create a system that is not just intelligent and effective but also naturally sustainable. By 2050, the long-term goal should be to create AI-guided transportation systems that are robust, flexible, and sustainable. To create a framework where AI serves as a catalyst for a global mobility network that is environmentally friendly and meets the requirements of all citizens, researchers, policymakers, and industry must work together.

Artificial intelligence can be instrumental in streamlining the processes in electric and hybrid vehicles by managing the battery, predicting health performance, and smart charging (Xiong et al., 2018). The usage patterns could be examined with the help of AI systems, which could lead to the optimization of battery life, energy consumption, and increase operational safety.

Moreover, it is possible with the help of AI to provide integration of vehicles with power grids, which means that electric cars can communicate dynamically with the power grids (Kempton & Tomic, 2005; Lopes et al., 2010). The ability reinforces the integration of renewable energy and grid stability as the supply and demand are balanced on a real-time basis. Integration of AI, transport, and energy firmly is one of the vital pillars of sustainability in urban development (Lopes et al., 2010).

2.8 Case Studies of AI-Driven Sustainable Transportation

A number of practical applications can demonstrate how AI in the context of sustainable mobility works. AI-enhanced adaptive traffic light system systems do this in Singapore, where real-time adjustments of intersection timing resulted in observable congestion and emissions reductions (Shladover, 2017). In China, AI-powered transportation systems will streamline bus and metro services by predicting passenger demand, enhancing the efficiency of the service delivery (Batty, 2013). The cities of Europe have already installed AI-controlled electric vehicle charging systems, which are connected to renewable energy sources, drastically decreasing the peak electricity consumption and carbon intensity (IEA, 2024).

3. ANALYSIS

Using data from 1995 to 2025, Figure 1 shows a network visualization of keywords associated with Sustainable AI Solutions for Transport Systems. As the largest nodes in the network, the graphic demonstrates that the most prevalent themes are grouped around artificial intelligence, deep learning, and generative artificial intelligence. The red cluster emphasizes the ethical, industrial, and trust-related aspects of AI and is mostly linked to ideas like trust, datasets, value alignment, and industry 4.0. The blue cluster focuses on technical elements like transfer learning, deep learning, and attention mechanisms, as well as how they might be used in fields like medication development and climate change. Terms like adult, treatment outcome, animals, schizophrenia, and health care are all connected by the green cluster, indicating a close relationship between AI research in sustainability and its uses in the biological and medical sciences. Emerging interdisciplinary links are shown by smaller nodes like the food industry, fintech, and psychological models. All things considered, the figure shows that AI in transportation is not a separate subject but rather is connected to more general discussions about sustainability, healthcare, and industrial innovation.

Figure 1. Bibliometric Analysis on Sustainable AI Solution for Transport System with data from Lens and made using VOSViewer

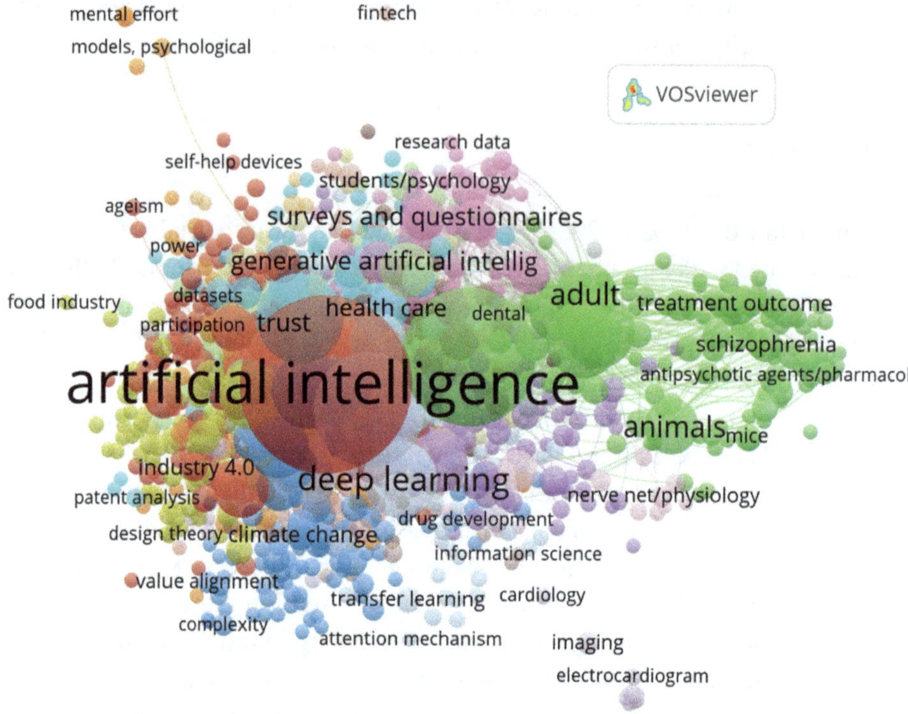

An overlay depiction of the temporal progression of research themes from 1995 to 2025 is presented in Figure 2. The early adoption of AI in healthcare and cognitive sciences is reflected in earlier research, which is represented by purple and blue tones (2018–2020) and focused on topics including psychological models, mental effort, cardiology, and animal studies. The emphasis moved to deep learning, attention mechanisms, transfer learning, and climate change between 2020 and 2022, suggesting a movement toward issues that are more technical and application-oriented. As explainability, human-AI interaction, and multidisciplinary applications gain prominence, more recent research—highlighted in yellow (2022–2024)—focuses on generative artificial intelligence, surveys and questionnaires, fintech, databases, and trust. Crucially, this new wave of industrial automation, environmental accountability, and ethical AI deployment includes the emergence of sustainable AI in transportation networks. This diagram illustrates the field's dynamic development from basic research to sophisticated, domain-specific applications.

Figure 2. Bibliometric Analysis on timeline of keywords on Sustainable AI Solution for Transport System with data from Lens and made using VOSViewer

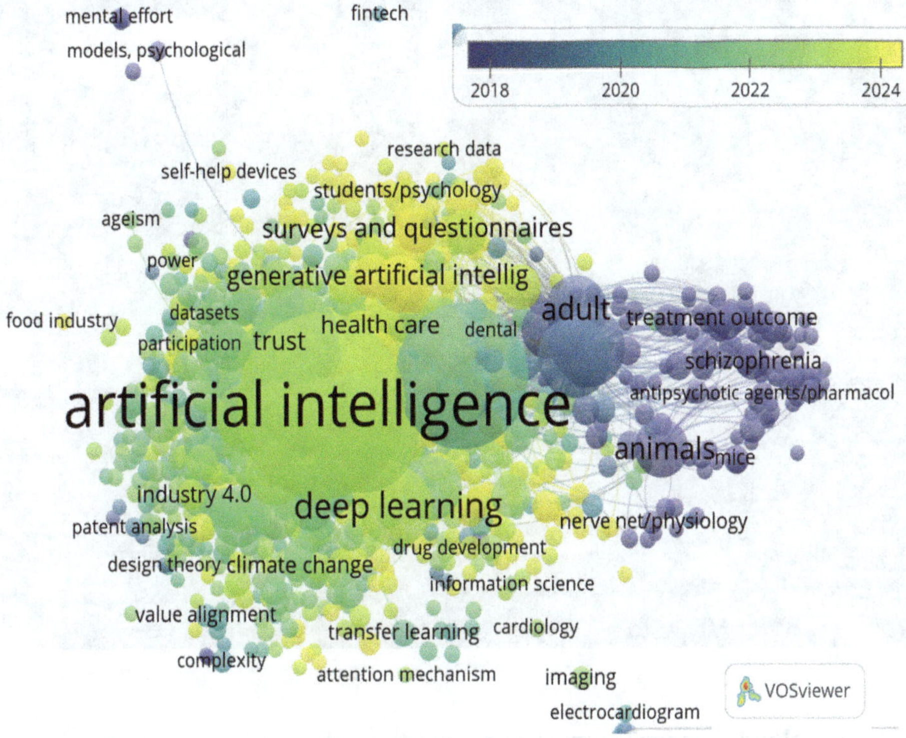

Brighter regions indicate areas of increased research concentration in the density visualization of the term co-occurrence network shown in Figure 3. Deep learning and artificial intelligence have the most intensive areas, demonstrating their importance to the subject. Strong research effort at the nexus of artificial intelligence, sustainability, and industrial transformation is indicated by the closely related hotspots of heath care, generative artificial intelligence, climate change, and industry 4.0. Active research in healthcare applications and ethical considerations is indicated by the moderately dense areas that surround trust, surveys and questionnaires, adults, animals, and treatment outcomes. Emerging but lesser study fronts are suggested by peripheral and less intense areas like the food business, fintech, ageism, and psychological models. Although transport-specific sustainable AI is still in its infancy, the density visualization supports the idea that it is closely related to major AI research areas like industrial applications, mitigating the effects of climate change, and innovative healthcare.

Figure 3. Bibliometric Analysis on density of keywords of Sustainable AI Solution for Transport System with data from Lens and made using VOSViewer

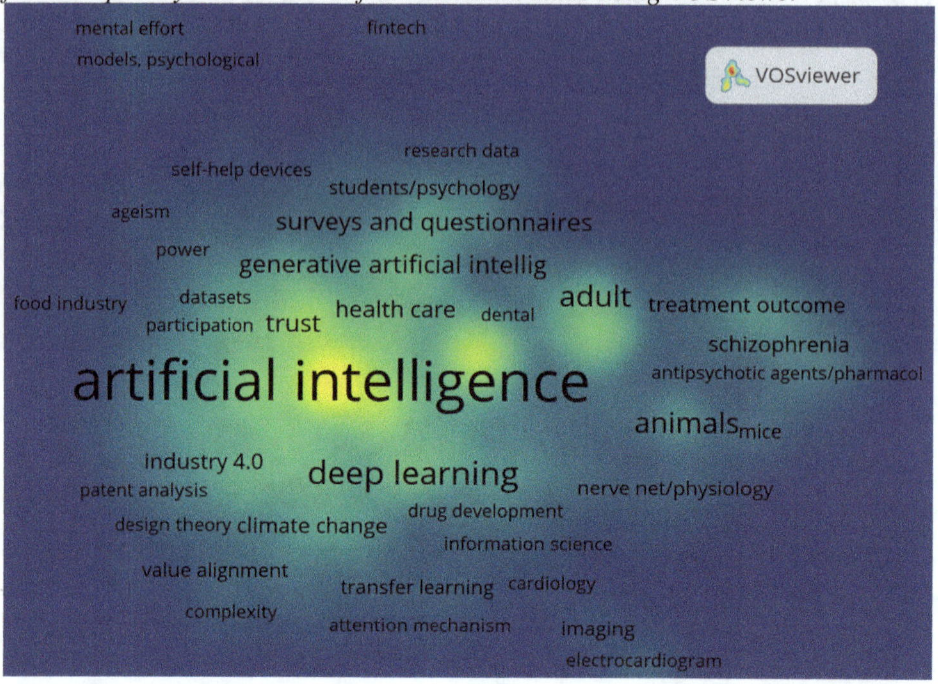

A word cloud showing the most common keywords in articles about sustainable AI solutions for transportation systems is shown in Figure 4. "Artificial Intelligence," "intelligence," "learning," "AI," "data," "health," "ethics," "technology," "education," "digital," "medical," and "machine" are among the most used terms. Learning-based models are fundamental to developments in this subject, as seen by the use of terminology like "Artificial Intelligence" and "learning" (including "machine learning" and "deep learning"). Words like "ethics," "privacy," "responsibility," "explainable," and "governance" are used frequently, which highlights the need for transparent and reliable AI frameworks and demonstrates a serious and growing concern for the ethical application of AI.

Furthermore, the use of words like "health," "healthcare," and "medical" implies a significant cross-disciplinary integration, showing that AI in transportation assists allied industries like public health and emergency response in addition to mobility. Terms like "sustainability," "network," "data," and "decision" indicate a heavy focus on environmentally conscious and data-driven systems that support the global Sustainable Development Goals (SDGs). Other words like "language," "model," "allocation," and "risk" draw attention to how crucial resource allocation, risk mitigation, and natural language processing are when designing AI systems for transportation. The topic is extremely diverse, combining technical innovation, ethical considerations, and policy engagement in the goal of intelligent and sustainable transportation solutions, as the keyword map illustrates overall.

Figure 4. Word cloud of keywords on Sustainable AI Solution for Transport System

The development of research interest in sustainable AI applications in the transportation industry is illustrated by Figure 5, which shows the temporal growth in scholarly publications from 1995 to 2025. Publications were continuously low between 1995 and 2016, with yearly production hardly ever exceeding 20 documents. But from about 2017, there is a noticeable uptick in research activity that goes hand in hand with the wider use of AI technology and the global expansion of smart mobility projects. After 2020, this trend picks up speed, with publication counts increasing substantially from about 300 in 2020 to over 2,100 in 2025—a seven-fold rise in only five years.

This surge was probably caused by a number of factors, such as the increasing use of AI in logistics, smart cities, and urban mobility systems; the increased emphasis on sustainability and climate resilience in transportation infrastructure; and the pressing need for automated and equitable transportation solutions during and after the COVID-19 pandemic. This high gradient indicates that academic, governmental, and industry players are paying close attention to sustainable AI in transportation, which is becoming a prominent interdisciplinary study topic. In addition to reflecting rising research output, the figure's exponential rise indicates that this field will continue to gain speed as countries around the world give priority to intelligent and net-zero transportation solutions.

Figure 5. Publication trend on research on Sustainable AI Solution for Transport System

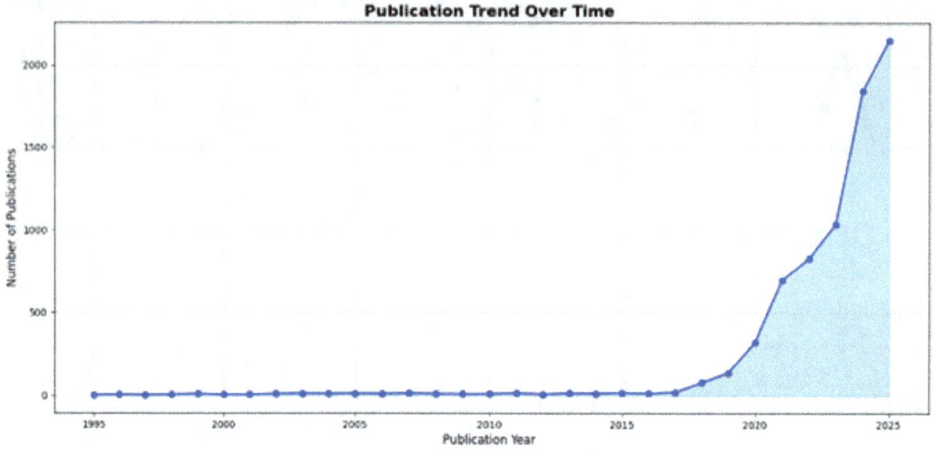

A worldwide summary of publishing activity by nation is shown in Figure 6, indicating which countries are at the forefront of sustainable AI research for transportation systems. With more than 650 published works, the United States is the most active contributor and is shown in the brightest yellow. The nation's strong university infrastructure, substantial financing for AI research, and the active involvement of federal agencies and tech companies are all factors contributing to its leadership position. High publishing quantities are also seen in China and India, highlighting their quickly expanding AI ecosystems and national emphasis on transportation infrastructure and smart city projects.

The United Kingdom, Germany, France, and the Netherlands are among the European nations that are heavily highlighted. European Union funding mechanisms like the Green Deal and Horizon Europe, which place a high priority on sustainable digital transformation, frequently support their robust research presence. With moderate document counts, emerging economies like Brazil, South Africa, and Indonesia may be seen on the map, indicating their growing interest in and participation in AI transportation advances. Large swaths of Africa, sections of South America, and Central Asia are still underrepresented, though, which may indicate differences in research capacity, access, or priorities.

Although sustainable AI in transportation is acknowledged as a global necessity, research efforts are still mostly focused in North America, Europe, and East Asia, as seen by the geographic distribution depicted in this image. In order to ensure that AI-driven transportation breakthroughs are fair and advantageous for all people worldwide, it may be possible to close this gap through international cooperation, technology exchange, and open-access research platforms.

Figure 6. Countries' graph based on their research and publication on Sustainable AI Solution for Transport System

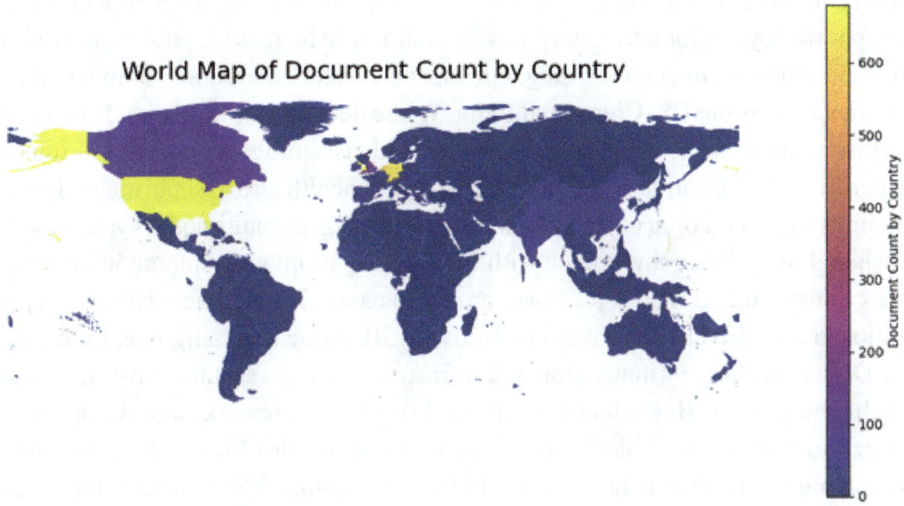

4. CONCLUSION

To sum up, this chapter has offered a comprehensive analysis of sustainable AI solutions for transportation systems, highlighting the critical role AI plays in tackling the complex issues facing the industry. The story demonstrates a paradigm shift toward proactive, data-driven mobility, from the introduction's summary of the economic and social significance of transportation in the face of urbanization, pollution, and infrastructure deterioration to the literature review's in-depth exploration of AI applications. Important domains like traffic management, which uses deep reinforcement learning to predict congestion and provide adaptive signals, show observable drops in emissions and fuel usage. While autonomous vehicles, driven by computer vision, sensor fusion, and V2X communication, promise previously unheard-of safety advantages by reducing human error, predictive maintenance and operational automation further improve efficiency by prolonging asset lifespans and avoiding disruptions. AI's frontier applications are demonstrated by the investigation of new systems like the Hyperloop, which optimize energy use, routing, and safety in fast-moving, low-friction environments. However, sustainability goes beyond technology; in order to guarantee equal benefits, ethical factors including data privacy, algorithmic bias, and inclusive design are crucial. The necessity for strong frameworks to get beyond obstacles and stop inequality from getting worse is highlighted by challenges that are technical (such as data integration and cybersecurity), infrastructural (expensive), regulatory (ambiguities surrounding responsibility), and sociological (e.g., job displacement, public trust). These conclusions are supported by bibliometric analysis, which shows an increase in publications after 2017 with hotspots in deep learning, ethics, and climate integration, dominated by countries such as the US, China, and India. While density and word clouds validate interdisciplinary links, stressing governance and risk mitigation, temporal trends demonstrate the transition from fundamental AI in healthcare to sophisticated, sustainability-focused applications. In the future, intelligent multimodal systems will be ushered in by AI's convergence with IoT and 5G, promoting smooth integration across transportation modes for lower emissions and improved accessibility. AI is positioned as a driver for net-zero mobility by 2050 due to its alignment with the UNSDGs, especially 9 (innovation and infrastructure), 11 (sustainable cities), and 13 (climate action). But achieving this goal requires teamwork: interdisciplinary research to improve scalable models, harmonizing policies for moral application, and community involvement to create. In the end, sustainable AI in transportation represents a dedication to social justice, environmental conservation, and robust infrastructure, going beyond efficiency. Stakeholders may use AI to build flexible, low-carbon ecosystems that satisfy demands around the world by emphasizing comprehensive integration over compartmentalized solutions. In order to close research gaps and guarantee that AI-driven transportation not only innovates but also improves societies around the world, future work should concentrate on real-world pilots, bias audits, and international standards. This chapter acts as a fundamental call to action, advocating for increased funding for ethical AI in order to create a more connected and environmentally friendly future. Artificial intelligence is one of the basic technologies of the future of sustainable transportation. It will create smarter traffic management, autonomous automobile, and electric mobility and energy integration with the help of AI. The success of these technologies in the long run, however, is subject to the establishment of well-built ethical principles, open governance systems, and non-discriminatory deployment culture. With the proper alignment of AI innovation with global sustainability objectives and human-oriented policy-making, transportation systems will be able to become safer, cleaner, and more equitable to help urban and regional development to be more resilient.

REFERENCES

Abduljabbar, R., Dia, H., Liyanage, S., & Bagloee, S. A. (2019). The role of artificial intelligence in traffic signal control: A systematic review. *Journal of Transport and Land Use, 12*(1), 163–194. DOI: 10.5198/jtlu.2019.1408

Al-Hussaini, S., Al-Shehri, H., & Al-Zahrani, A. (2021). A review of AI-based solutions for smart parking management in urban areas. *Applied Sciences (Basel, Switzerland), 11*(13), 6100. DOI: 10.3390/app11136100

Bhatnagar, K., & Sastry, C. (2020). Artificial intelligence and machine learning for sustainable transportation: A review. *Journal of Sustainable Development of Energy. Water and Environment Systems, 8*(3), 478–499. DOI: 10.13044/j.sdewes.d7.0270

Bonnefon, J., Shariff, A., & Rahwan, I. (2016). The social dilemma of autonomous vehicles. *Science, 352*(6293), 1573–1576. DOI: 10.1126/science.aaf2654

Chen, Z., Guo, S., & Li, Y. (2020). Sustainable traffic signal control using deep reinforcement learning with an attention mechanism. *IEEE Transactions on Intelligent Transportation Systems, 21*(9), 3788–3798. DOI: 10.1109/TITS.2019.2934094

Dhivya, M., & Gowthami, S. (2020). AI-driven traffic management system for smart cities. *International Journal of Computer Engineering and Technology, 11*(4), 161–169. DOI: 10.34218/IJCET.11.4.2020.019

Floridi, L., Cowls, J., Beltrametti, M., Chatila, R., Chazerand, P., Dignum, V., Luetge, C., Madelin, R., Pagallo, U., Rossi, F., Schafer, B., Valcke, P., & Vayena, E. (2018). AI4People—An Ethical Framework for a Good AI Society: Opportunities, risks, principles, and recommendations. *Minds and Machines, 28*(4), 689–707. DOI: 10.1007/s11023-018-9482-5

Gao, S., Ma, X., & Li, H. (2020). A deep learning-based model for predicting traffic congestion in urban areas. *Journal of Big Data, 7*(1), 1–18. DOI: 10.1186/s40537-020-00336-7

Garg, N., & Sharma, V. (2022). A review of AI and IoT-based solutions for smart and sustainable cities. *International Journal of Sustainable Engineering, 15*(1), 1–13. DOI: 10.1080/19397038.2022.2030113

Ghasemi, H., & Zarrabi, M. (2022). Optimizing urban freight delivery using a multi-objective meta-heuristic algorithm: A case study with electric vehicles. *Journal of Cleaner Production, 357*, 131976. DOI: 10.1016/j.jclepro.2022.131976

Goudarzi, G., & Mohammadi, A. (2021). AI applications in Hyperloop systems: A review on safety, efficiency, and sustainability. *Journal of Transportation Engineering. Part A, Systems*, *147*(9), 04021045. DOI: 10.1061/JTEAPL.0000620

Guo, Y., Shi, F., & Li, C. (2021). A comprehensive review of AI applications for sustainable transportation systems. *Sustainability*, *13*(24), 13867. DOI: 10.3390/su132413867

Hofstetter, G., Ma, Z., & Sgouridis, S. (2019). The role of Hyperloop in a sustainable transportation future: A techno-economic and environmental assessment. *Transportation Research Part D, Transport and Environment*, *69*, 312–327. DOI: 10.1016/j.trd.2019.02.008

Huang, Y., Xu, J., & Zhang, H. (2019). Reinforcement learning for energy-efficient train operation in high-speed rail. *IEEE Transactions on Intelligent Transportation Systems*, *20*(9), 3469–3478. DOI: 10.1109/TITS.2018.2882252

Kempton, W., & Tomić, J. (2005). Vehicle-to-grid power fundamentals: Calculating capacity and net revenue. *Journal of Power Sources*, *144*(1), 268–279. DOI: 10.1016/j.jpowsour.2004.12.025

Li, G., Song, M., & Sun, Y. (2020). AI-based predictive maintenance for urban rail transit systems. *IEEE Transactions on Industrial Informatics*, *16*(3), 1957–1965. DOI: 10.1109/TII.2019.2929427

Liu, Y., Li, J., & Yang, B. (2021). A multi-agent reinforcement learning approach for green traffic signal control. *IEEE Transactions on Systems, Man, and Cybernetics. Systems*, *51*(8), 4991–5001. DOI: 10.1109/TSMC.2019.2946272

Lopes, J. P., Soares, F. J., & Almeida, P. M. R. (2010). Integration of electric vehicles in the electric power system. *Proceedings of the IEEE*, *99*(1), 168–183. DOI: 10.1109/JPROC.2010.2066250

Luo, J., & Zhang, W. (2019). Carbon emission reduction in a logistics network using an AI-based vehicle routing algorithm. *Journal of Cleaner Production*, *222*, 974–984. DOI: 10.1016/j.jclepro.2019.03.040

Mishra, S., & Garg, P. (2022). A deep learning approach for traffic flow prediction in a smart city environment. *Journal of Ambient Intelligence and Humanized Computing*, *13*(5), 2379–2391. DOI: 10.1007/s12652-021-03004-8

Mohanty, S., Kumar, S., & Dash, R. R. (2022). AI-driven smart infrastructure for sustainable transportation in smart cities. In Dash, R. R., & Mohanty, S. S. (Eds.), *Smart Cities and Urban Development* (pp. 235–256). Wiley., DOI: 10.1002/9781119792015.ch10

OECD. (2023). AI in transportation systems policy frameworks. IEA. (2024). AI for energy and mobility integration.

Quiroga, R. Q. (2012). Borges and memory. In *The MIT Press eBooks*. DOI: 10.7551/mitpress/9396.001.0001

Saadi, A., Abghour, N., Chiba, Z., Moussaid, K., & Ali, S. (2023). A survey of Reinforcement and Deep reinforcement learning for Coordination in Intelligent Traffic Light Control. *Reseach Square*. DOI: 10.21203/rs.3.rs-3225879/v1

Sarkar, D., & Pal, S. (2020). AI-enabled fleet management for carbon footprint reduction in last-mile delivery. *Procedia Computer Science*, *171*, 1642–1651. DOI: 10.1016/j.procs.2020.06.168

Shah, S., Khan, K., & Ahmad, N. (2021). AI-based predictive maintenance of autonomous vehicles for enhanced safety and reliability. *Sensors (Basel)*, *21*(2), 438. DOI: 10.3390/s21020438

Shladover, S. E. (2017). Connected and automated vehicle systems: Introduction and overview. *Journal of Intelligent Transport Systems*, *22*(3), 190–200. DOI: 10.1080/15472450.2017.1336053

Teng, J., & Guo, Q. (2020). AI-enabled dynamic pricing for ride-sharing to reduce traffic congestion and carbon emissions. *European Journal of Operational Research*, *287*(2), 652–663. DOI: 10.1016/j.ejor.2020.04.053

Transforming our world: the 2030 Agenda for Sustainable Development | Department of Economic and Social Affairs. (n.d.). https://sdgs.un.org/2030agenda

UNCTAD. (2023). Digital transformation and transport equity.

Vasan, A., & Balamurugan, K. (2021). Autonomous vehicles and their impact on sustainable transportation. In Pal, S. K. (Ed.), *Intelligent Transportation Systems* (pp. 201–220). Springer. DOI: 10.1007/978-981-15-7798-9_10

Wang, K., Zhang, Q., & He, G. (2021). Sustainable mobility in smart cities: A review of AI-enabled solutions. *Sustainable Cities and Society*, *66*, 102657. DOI: 10.1016/j.scs.2021.102657

Wang, Y., Li, Q., & Sun, Z. (2018). Artificial intelligence in intelligent transportation systems: A review of applications, challenges, and opportunities. *IEEE Access : Practical Innovations, Open Solutions*, *6*, 62968–62985. DOI: 10.1109/ACCESS.2018.2877990

Xiong, R., Zhang, Y., Wang, J., He, H., Peng, S., & Pecht, M. (2018). Lithium-Ion battery health prognosis based on a real battery management system used in electric vehicles. *IEEE Transactions on Vehicular Technology*, *68*(5), 4110–4121. DOI: 10.1109/TVT.2018.2864688

Zhang, J., Zhang, C., & Zhang, Y. (2020). AI-driven logistics and supply chain management for sustainable transportation. *Transportation Research Part E, Logistics and Transportation Review*, *138*, 101968. DOI: 10.1016/j.tre.2020.101968

Zou, Y., Li, M., Shi, M., & Yang, B. (2021). Predictive maintenance for electric vehicles using machine learning: A case study. *Energy*, *224*, 120150. DOI: 10.1016/j.energy.2021.120150

ADDITIONAL READING

Chen, H., & Rakha, H. (2023). Integrated optimization of vehicle trajectories and traffic signal timings. *Proceedings of the 7th International Conference on Vehicle Technology and Intelligent Transport Systems*, 27–34. DOI: 10.5220/0011709900003479

Jagatheesaperumal, S. K., Bibri, S. E., Huang, J., Rajapandian, J., & Parthiban, B. (2024). Artificial intelligence of things for smart cities: Advanced solutions for enhancing transportation safety. *Computational Urban Science*, *4*(1), 10. Advance online publication. DOI: 10.1007/s43762-024-00120-6

Malaj, A., & Muka, E. (2023). AI Lights and Shadows: Revolutionizing the world. *FORUM a+P Interdisciplinary Journal of Architecture and Built Environment, 27*, 84–91. DOI: 10.37199/f40002712

Rumin, R. (2024). Utilizing artificial intelligence for energy-efficient route planning in hyperloop low-pressure capsule transit: A study in alignment with sustainable development goals. *Scientific Papers of Silesian University of Technology Organization and Management Series*, *204*. Advance online publication. DOI: 10.29119/1641-3466.2024.204.32

Wei, H., Xu, N., Zhang, H., Zheng, G., Zang, X., Chen, C., Zhang, W., Zhu, Y., Xu, K., & Li, Z. (2019). CoLight: Learning Network-level Cooperation for Traffic Signal Control. In *International Conference on Information and Knowledge Management*. DOI: 10.1145/3357384.3357902

KEY TERMS AND DEFINITIONS

Artificial Intelligence (AI): It is the process by which machines mimic human intelligence to carry out activities like learning, reasoning, and making decisions.

Autonomous Cars: These are automobiles that can drive themselves without assistance from a driver thanks to sensors and artificial intelligence.

Hyperloop: It is a proposed high-speed transit system that moves pods at nearly airplane speed utilizing magnetic levitation and low-pressure tunnels.

Machine Learning (ML): It is a branch of artificial intelligence that allows systems to recognize patterns in data and enhance performance without the need for explicit programming.

Safety: It is the state of being shielded from danger, injury, or mishaps, especially in settings, systems, and modes of transportation.

Sustainability: Meeting current requirements without sacrificing the capacity of future generations to meet their own.

Traffic Management: It is the process of organizing, observing, and regulating traffic flow to guarantee sustainable, effective, and safe transportation.

Transportation: Moving people or things from one location to another by air, sea, train, or road.

Chapter 6

Automation and Digitalization in the Automotive Industry:

Employment Impacts From the Case of Toyota in Mexico

Jorge Carrillo
https://orcid.org/0000-0002-8533-2797
El Colef, Mexico

Adriana Martínez
https://orcid.org/0000-0003-0639-4007
ENES-León UNAM, Mexico

ABSTRACT

This study analyzes the impact of automation and digitalization on employment through the lens of Skill-Biased Technological Change (SBTC), using the Toyota Guanajuato plant (TMMGT) as a case study. Using a qualitative methodology, the study identified an organizational strategy centered on internal relocation, continuous training, and upward mobility. This has allowed for the integration of new technologies without generating labor disruptions. The findings show that the employment effects of automation depend less on the technology itself and more on the organizational decisions that accompany its implementation. The TMMGT case exemplifies how a corporate culture focused on continuous improvement and human development can mitigate the exclusionary effects predicted by SBTC theory, enabling a more inclusive technological transition. This study contributes to the

DOI: 10.4018/979-8-3373-4495-9.ch006

literature on digital transformation and employment in emerging economies. Future research should explore how organizational practices and institutional frameworks mediate the effects of technological change on work.

NOTE

In collaboration with Ana Karina Hernández, El Colegio de la Frontera Norte, Mexico.

INTRODUCTION

The impact of technological change on employment has been a subject of debate since the First Industrial Revolution, beginning with the introduction of the steam engine, and later with the arrival of electricity and computing in production processes. Today, we are as the cusp of the so-called Fourth Industrial Revolution, driven by Industry 4.0 technologies such as automation, robotics, artificial intelligence (AI), and the Internet of Things. These innovations are not only reshaping production systems but also redefining work organization and labor dynamics, intensifying concerns about workforce displacement and the polarization of skills (Schwab, 2016; OECD, 2019).

In this context, the emergence of intelligent, autonomous, and interconnected technologies constitutes a socio-technical ecosystem that transforms the role of human labor. While some perspectives emphasize their labor-substituting effects (Kamaruddin *et al.*, 2013; Eurofound, 2018), others highlight the generation of new occupations requiring advanced digital, cognitive, and technical competencies (Autor *et al.*, 2020; Anzolin, 2021).

From the perspective of the Skill-Biased Technological Change (SBTC) framework, technological innovations do not uniformly eliminate employment but rather reshape the structure of labor demand, favoring workers with higher skills and reducing demand for routine-based occupations (Katz & Murphy, 1992; Berman, Bound, & Griliches, 1994; Autor, Levy, & Murnane, 2003; Acemoglu, 2002; Goldin & Katz, 2008). They favor workers with more advanced technical and cognitive skills while reducing the demand for low-skilled tasks (Katz & Autor, 1999; Autor, 2015). From this point of view, automation, robotization, and digitalization do not necessarily lead to a net reduction in employment but rather a transformation in the types of labor profiles required. As companies adopt advanced technologies—such as collaborative robotics, AI, and cyber-physical systems—the demand increases for workers capable of operating, maintaining, and optimizing these systems, to the detriment of occupations based on routine tasks. This perspective challenges the traditional narrative that links automation with unemployment and invites us to consider alternative scenarios for the transformation of employment and required skills.

The automotive industry, often described as "the industry of industries," provides an ideal context to analyze technological and organizational transformation due to its extensive and complex value chains, its integration of diverse upstream and downstream sectors, and its historical role as a driver of innovation and industrial upgrading (Womack, Jones, & Roos, 1990; Humphrey & Memedovic, 2003; Sturgeon, Van Biesebroeck, & Gereffi, 2008). It has historically pioneered production models such as Fordism, Lean Production, and now Industry 4.0 (Boyer & Freyssenet, 2003; Kagermann, 2013). Moreover, its role as a driver of technological innovation directly impacts the global labor market, reducing traditional jobs while creating new professional profiles in areas such as data analysis, software development, and systems integration (Carrillo *et al.*, 2023).

The main objective of this article is to understand how automation and digitalization processes are being implemented in the automotive industry of an emerging economy like Mexico, and what effects they have on employment and work organization. To this end, we analyze the case of the Toyota Motor Manufacturing plant in Guanajuato (TMMGT). Based on fieldwork conducted in three stages—in 2021, 2023, and 2025—including guided visits and interviews with authorized personnel, we obtained direct information on the technological transformations underway. The research question guiding this study is: Are automation and digitalization generating unemployment in the automotive industry? What types of skills are being affected?

The article is structured into six sections in addition to this introduction. The following section presents a literature review on automation, digitalization, and Industry 4.0, along with the main theoretical approaches used to study their impact on employment in the second section. Next, the methodology is described, and the studied plant is briefly contextualized. The fourth section presents the case study findings and discusses them through the lens of the SBTC approach. The fifth section shows the discussion and finally, the article concludes with a reflection on the impact of these technologies at TMMGT and the key role of organizational management in this transformation.

1. INDUSTRY 4.0 IN THE AUTOMOTIVE SECTOR: AUTOMATION, DIGITALIZATION, AND EMPLOYMENT TRANSFORMATION

The automotive industry is undergoing a critical phase of transformation driven by technologies grouped under the concept of Industry 4.0 (I4.0). Its key pillars include the Internet of Things, AI, collaborative robotics, big data analytics, and digital platforms. These technologies are reshaping production systems by blurring the boundaries between physical and digital environments (Andreoni & Anzolin, 2025), as they enable machines, systems, and people to be connected in real time, facilitating faster and more accurate decision-making. While not entirely new processes, their current scope has reached unprecedented complexity and scale, with significant impacts on production organization, employment, and global supply chains.

In automotive plants, a new dynamic is emerging between workers and automated systems, enhancing the flexibility and productivity of assembly lines. The adoption of these technologies responds not only to demands for efficiency and quality (Autor, 2015), but also to global pressures linked to vehicle electrification (Bohnsack, Pinkse & Kolk, 2014), competitiveness (OECD, 2020), and national and international policies (UNIDO, 2022).

Although there is no single definition of I4.0, some authors describe it as a form of advanced connectivity throughout the entire value chain, based on autonomous learning and digital interaction (Smit *et al.*, 2016; Sukhodolov, 2019). According to Axis (2019), there are at least 19 key technologies associated with I4.0, with the highest levels of adoption found in multinational industries, particularly in manufacturing (Carrillo *et al.*, 2023).

In this context, we also identify transversal technologies related to I4.0 whose origins date back to the Third Industrial Revolution: automation, robotization, and digitalization. While not recent, these technologies have been fundamental to the development of I4.0. According to Anzolin (2021), automation began in the 1960s with the introduction of robotic arms in industrial production and intensified with computer numerical control and advanced robotics. The author identifies three major technological trends that have impacted manufacturing: mechanization, automation, and robotization, which have evolved sequentially, with each stage paving the way for the next. According to Richard (2005) and Kamaruddin *et al.* (2013), mechanization refers to the replacement of human labor by mechanical work; automation, to the replacement of human "thinking" through programmed systems; and robotization involves greater complexity in task execution, as programmed machines can modify the type and intensity of tasks with growing self-regulation capabilities (Anzolin, 2021). Digitalization, in turn, represents an even broader process, beginning with the conversion of analog to digital data (digitization) and culminating in interconnected networks of machines, software, and databases that allow real-time control and optimization of processes (Hirsch-Kreinsen & Hompel, 2017; Anzolin, 2021).

In summary, automation, robotization, and digitalization are interrelated, though they serve different functions and have distinct scopes within industrial and organizational processes: digitalization provides the data, connectivity, and technological infrastructure that enables automated processes; automation is linked to task programming and control via mechanical systems; and robotization constitutes the physical materialization of those tasks through programmable devices. Generally, digitalization is oriented toward software, automation toward hardware, and robotization is a convergence of both. While each respond to different technical logics, in practice, they are implemented in an integrated manner as part of the same productive ecosystem.

For this reason, the present study addresses them in an integrated way, recognizing that automation and robotization are the most visible and observable processes in the analyzed case, while digitalization operates as the enabling infrastructure that supports and connects these systems. Thus, the empirical analysis focuses on the concrete manifestations of automation and robotization at the plant, without losing sight of the cross-cutting role of digitalization in organizational and productive transformation.

2. UNDERSTANDING EMPLOYMENT IMPACTS: THEORY AND EVIDENCE FROM MEXICO

The impact of technology on employment has long been a major concern, particularly for economists and sociologists. Various theories and approaches have emerged to understand and predict its effects on the labor market. From a determinist perspective, Keynes (1930) predicted "technological unemployment" by the 2030s. Later, other frameworks attempted to explain how technological progress affects employment. One of the most influential was the concept of creative destruction, developed by Schumpeter (1942), who argued that while technological change destroys existing sectors and jobs, it also creates new industries, products, and forms of work. In this view, innovation acts as a transformative force that drives economic growth but does so through a disruptive process that does not guarantee displaced workers will automatically benefit.

The SBTC approach emerged in the 1980s in the context of rising wage inequality in developed countries (Katz & Murphy, 1992). It predicts that the demand for "skilled" jobs is increasing relative to "unskilled" ones (Goos & Manning, 2007). As a complement to this approach, Autor, Levy, and Murnane (2003), in a study conducted in the United States, proposed that technology replaces human labor mainly in routine tasks, but not in non-routine ones—regardless of skill level, qualification, or job position. Similarly, Goos and Manning (2007) argued that routinization is a better explanatory factor for job polarization. Their study, conducted in the United Kingdom between 1975 and 1999, showed that technology fostered the growth of both high- and low-skilled jobs while reducing those in the middle—a phenomenon known as task-based technological change. For example, a highly skilled data analyst may have certain routine functions automated through machine learning algorithms, while a maintenance technician performing non-routine tasks may increase in value by overseeing that same technology. This introduces an additional dimension to the SBTC framework by focusing on the nature of the task itself.

Recent studies continue to observe this trend. One such example is the research by Omar, Mohd, and Syama (2025), conducted in Malaysian manufacturing firms between 2012 and 2023, which highlights the dual role of automation: on the one hand, it boosts productivity; on the other, it intensifies labor market polarization.

While both classical and contemporary studies (Keynes, 1930; Schumpeter, 1942; Autor-Levy-Murnane, 2003; Omar, Mohd & Syama, 2025) describe how new technologies may generate technological unemployment or polarize the labor market, these dynamics take on very specific forms depending on the productive context and institutional characteristics of each region.

In Mexico, the automotive industry has been widely studied, particularly from economic perspectives focused on exports, productivity, value-added, and integration into global value chains (Álvarez & González Marín, 2017; Lampón, *et al.* 2018; Moreno Brid *et al.*, 2021; Arteaga, 2020; Covarrubias, 2020). In parallel, labor studies have focused on working conditions, union dynamics, and workforce composition (Middlebrook, 1988; De Buen, 2011; Covarrubias & Bouzas, 2016; García-Jiménez *et al.*, 2021). More recently, some studies have begun to analyze the unequal adoption of automation technologies among original equipment manufacturers (OEMs) (Bensusán & Gómez, 2017; Morales & López, 2018; Reyes et al., 2018). However, studies examining the concrete implementation of Industry 4.0 technologies and their effects on labor dynamics inside plants—especially from a qualitative perspective—are still limited.

The automotive sector in Mexico —which accounts for 18.3% of manufacturing GDP and employs over one million workers—has undergone phases of mechanization, automation, and more recently, robotization and I4.0. In fact, of the 5,869 robotic units installed in the country in 2023, 69% were in automotive plants (IFR, 2024). The Mexican automotive industry is highly dynamic; globally, it ranks 5th in vehicle production and 4th in auto parts manufacturing (OICA 2022; AMIA 2024). Between January 2020 and November 2024, 16,920,646 light vehicles were produced, representing a 24% increase over that period (Méndez, 2025). Thus, the Mexican automotive industry is a key case for observing automation, robotization, and digitalization in emerging markets under the I4.0 framework.

3. METHODOLOGY

This study employs a qualitative, longitudinal case study of Toyota Motor Manufacturing Guanajuato (TMMGT) to examine how automation, robotization, and digitalization shape work organization and employment in Mexico's automotive sector. Unlike macro-level or policy-oriented approaches, the research emphasizes the micro-dynamics of technology adoption within a newly established plant designed from the outset with advanced manufacturing capabilities.

The research design combined documentary review, non-participant observation, and semi-structured interviews with employees, managers, and public officials directly engaged in Industry 4.0 processes, innovation projects, and talent development. Data collection unfolded in three stages between 2021 and 2025, generating 23 interviews complemented by site visits and corporate documentation. This longitudinal strategy enabled temporal triangulation, allowing the identification of both continuities and changes over time.

Data were coded thematically using a hybrid strategy that combined deductive categories from Industry 4.0 and organizational change literature with inductive insights emerging from the fieldwork. Reflexivity and ethical safeguards were central throughout: informed consent and anonymity were ensured, while triangulation across multiple sources strengthened validity and preserved critical distance.

Overall, the methodology demonstrates how sustained qualitative engagement can capture the lived experience of digital transformation in manufacturing, highlighting the micro-foundations of technological change often overlooked in policy analyses.

4. RESEARCH RESULTS

4.1 Description of the Plant

Toyota Motor Manufacturing de Guanajuato (TMMGT) began operations in Apaseo El Grande, with an initial investment of 700 million U.S. dollars. The plant spans 620 hectares and 240,000 square meters of built area. Its location was chosen based on three key factors: efficient logistical access (by sea and rail), integration with the Bajío automotive cluster, and the availability of skilled technical labor.

TMMGT produces 100% of the Tacoma pickup trucks sold in North America and, along with the Tijuana plant, is the only facility in the world assembling this model. Between 2021 and 2024, annual production increased from 100,000 to 138,000 units, and the number of direct employees grew from 2,233 to 2,755. Seventy-two percent of components are produced locally, and over 90% of vehicles are exported to the United States and Canada.

The plant comprises the following areas: stamping, bodywork, welding, painting, assembly, and quality control. It also includes a training area and a water treatment plant that recycles 80% of the water used, with the remaining 20% repurposed for irrigation. The production of hybrid or electric versions of the Tacoma largely depends on market demand. In 2024, hybrid versions already accounted for 26% of total production (Rodríguez & Oropeza, 2024).

The organizational model is based on the Toyota Production System (TPS), which combines efficiency, quality, and employee participation. It is implemented through self-managed work cells, continuous training, and job rotation. The organizational structure is horizontal, with seven hierarchical levels, and most employees are affiliated with the SITIMM-CTM union. Several automotive plants in Guanajuato are unionized under SITIMM. Although formally affiliated with the CTM, the union retains a degree of autonomy and can be characterized as active, maintaining constructive relations with management, operating with transparency, and achieving significant contractual gains.

Since its opening, the plant has prioritized technical training and technological adaptation. Engineers from the Tijuana plant were transferred to accelerate the learning curve, and partnerships were established with local universities such as the Technological Institute of Celaya (ITC), the University of Guanajuato (UG), and the Autonomous University of Querétaro (UAQ). Through the Fresh Engineers program and collaborative I4.0 projects, young professionals have been integrated into areas such as robotics, mechatronics, and AI.

To ensure efficiency and just-in-time production, TMMGT relies on sixteen global suppliers, eleven of which operate within the industrial campus and employ 4,783 people. In total, the campus hosts 7,538 workers, of whom 63% are employed by supplier companies. While they share the same union as Toyota workers, their wages and benefits are slightly lower. It is worth noting that each unionized plant in Mexico's automotive sector exhibits differences in employment conditions. Rather than being organized under an industry-wide union, the sector is characterized by plant-level unions. As a result, financial arrangements, cost structures, and positions within the value chain shape the distinct employment conditions negotiated at each facility. Although such disparities may occasionally generate discontent among union members at a specific plant, this situation is largely normalized within the Mexican context. Consequently, labor turnover in the sector tends to be relatively high.

The wage structure at TMMGT includes nine levels, with hourly wages ranging from $1.60 to $3.60 USD, and an average of $2.40. Wage dispersion is low (a ratio of 2.2 between the highest and lowest levels). Although salaries are lower than those at other Toyota plants in North America, they exceed the regional average. Benefits include transportation, a shared cafeteria (with suppliers), wellness programs, and specific support for female workers, such as nutritional counseling during pregnancy, a lactation room, and union support for social security procedures.

At TMMGT, training is a central pillar and is structured in three stages: an initial five-day course (focused on quality, the TPS, and labor-related aspects), technical training at the "dojo" center located on the production line (lasting from two weeks to three months), and hands-on training on the line until autonomy is achieved. With the introduction of new technologies, Toyota has identified the strengthening of soft skills—such as creativity, teamwork, problem-solving, and resilience—as a key area. These skills complement technical competencies, which are acquired through courses and/or reinforced through projects such as the collaborative I4.0 initiatives developed in partnership with local universities.

4.2 Patterns of Automation and Technological Change (2021–2024)

The adoption of automation, robotization, and digital technologies at TMMGT has been deliberate and staged, with distinct rhythms across production areas. While the entire plant integrates elements of I4.0, the intensity of automation varies depending on the operational requirements, safety considerations, and potential productivity gains of each stage.

Table 1 summarises production volumes, employment levels, and robot deployment between 2021 and 2024. The data show a threefold increase in the number of robots and a simultaneous expansion of the workforce. This trajectory challenges deterministic interpretations equating automation with job elimination (Autor, 2015; Acemoglu & Restrepo, 2019), suggesting instead a complementary dynamic in which technology and labour evolve concurrently.

Table 1. Production, employment and robots in TMMGT, 2021-2024

Year	Vehicles production	Total employment	No. of robots	Robots by worker	Vehicles by worker
2021	100,000	2,233	127	0.06	44.8
2022	138,000	2,252	176	0.08	61.3
2023	123,000	2,717	347	0.13	45.3
2024	138,000	2,755	521	0.19	50.1

Source: Author's elaboration

Between 2021 and 2024, TMMGT underwent a profound transformation in its production structure, characterized by sustained growth in the number of industrial robots, an increase in employment, and a recovery in productivity. The number of installed robots grew by more than 300%, rising from 127 in 2021 to an estimated 521 in 2024. Robot density per worker increased from 0.06 to 0.19, reflecting intensive and progressive automation.

In parallel, the total workforce grew by 23.3%, from 2,233 to 2,755 employees, though not in a linear fashion. The most significant increase occurred in 2023 (20.6%), coinciding with a wave of technological expansion, followed by more moderate growth in 2024 (1.4%), which may indicate a stabilization phase following the expansion.

Labor productivity (vehicles per worker) showed dynamic behavior, peaking in 2022 (61.3), then dropping in 2023 (−26.1%), which may be attributed to a temporary mismatch between technology and training, followed by a recovery in 2024 (+10.6%), this dip may suggest better organizational adaptation and technological appropriation.

This set of data challenges the simplistic narrative that links automation with job destruction. At TMMGT, a complementarity between technology and skilled human capital is observed, confirming that technological implementation alone does not guarantee efficiency gains; its impact depends on the organizational structure, the type of tasks being automated, and the degree of technological appropriation by the workforce.

One of the main findings from the fieldwork at TMMGT is the gradual implementation of automation, robotization, and digitalization technologies in areas such as bodywork, stamping, and logistics, under a collaborative approach. Rather than adopting disruptive strategies, Toyota has embraced an evolutionary model centered on continuous improvement and the progressive integration of technological solutions developed in-house or in collaboration with local suppliers, avoiding "black boxes" and closed automated systems (Kunkel *et al.*, 2022; Bughin & Manyika, 2019). This digital transformation, aligned with the TPS, aims to empower workers through collaborative tools and is implemented through consensus-based changes supported by operational knowledge (Frank *et al.*, 2019; Kunkel *et al.*, 2022).

This gradual and collaborative approach to automation is also reflected in the current distribution of robots and digital technologies by area within the plant. The following table summarizes how human labor is combined with varying levels of automation depending on the specific characteristics of each process:

Table 2. Level of automation, digitalization and robotization by area in TMMGT (2021)

Area	Employees %	No. of robots	% of Robots	Worker by robots	Task characteristics	Technology	Level of automation
Stamping	5.6%	10	8%	12.5	Continuous pace, heavy load, routine tasks	12 robots	High
Press					Continuous pace, heavy load, routine tasks	Latest generation of technology	High
Body shop	17.8%	69	54%		High precision, physical risk	58 robots, 1 for training, 4 AGV, 1 Automatic conveyor	Very high
Painting	16.9%	44	35%	8.6	Controlled environments, chemical handling, high repetitiveness	Full automated	Very high
Welding (subarea)	—	—	—	5.8	Repetitive tasks, need for precision	50% automation*	High
Final assembly	32.6%	—	—	243	Fine-tuning, flexibility, intensive human intervention	Manual work, Pneumatic stations, support robots	Low
Quality control	13.1%	—	—	—	Sensory evaluation, human interpretation of errors	Visual and measurement tools	Low - Medium
Other (logistics, support)	13.9%	—	—	—	Indirect support, administration, and managmentn,	Mixed: digitalization, logistics	Variable

Source: Author's elaboration based on information from TMMGT (2021).

* In 2024, laser welding technology was introduced, improving the precision and efficiency of the process.

In 2021, the plant had 127 industrial robots, mainly concentrated in bodywork (54%), painting (35%), and stamping (8%), reflecting a selective automation strategy focused on processes requiring high precision, repetitiveness, or that involve physical risks. This aligns with the analysis of robot density per employee, which also shows differentiated patterns based on the nature of each area:

- Stamping: 1 robot for every 12.5 employees
- Painting: 1 robot for every 8.6
- Welding: 1 robot for every 5.8

- Final assembly: 1 robot for every 243

These data reveal a strong concentration of automation in technical areas like welding and painting, which, although they employ a smaller share of the total workforce, integrate a considerable number of robots. In contrast, final assembly—with a low level of automation—employs nearly one-third of the labor force, underscoring the central role of human labor in this stage of the production process.

This disparity in the robot-to-worker ratio does not indicate a lag but rather strategic decisions about where to automate and where to preserve human involvement. TMMGT prioritizes automation that complements human labor: only those tasks where machines add real value are automated, maintaining a balance between efficiency, safety, quality, and employment. This approach promotes a responsible and inclusive technological transition that integrates people into the process and considers both social and operational impacts.

This configuration reveals a selective automation strategy: physically demanding, repetitive, or high-risk tasks are prioritized for automation, while those requiring adaptability, judgment, or direct interaction retain high levels of human involvement. This means that rather than replacing people, technological implementation at TMMGT reconfigures work. Robots are concentrated in processes with low variability and a high degree of standardization, while stages requiring flexibility and responsiveness continue to rely on human judgment and intervention.

From an SBTC perspective, the pattern at TMMGT favors the retention and upskilling of workers whose roles involve non-routine, cognitive, or socially interactive tasks (Autor, 2015). Automation displaces repetitive manual activities but simultaneously raises the demand for competencies in troubleshooting, coordination, and process improvement.

Thus, the TMMGT case illustrates how robotization does not advance uniformly but is modulated by the technical nature of tasks—reinforcing the idea of technological change being biased by job function, not just by skill level. This approach also helps explain why the plant has managed to maintain high employment levels without mass layoffs, even while incorporating advanced technology

4.3 Organizational Drivers and Institutional Effects

At TMMGT, automation has been driven by motivations related to economic efficiency, improved occupational safety, and environmental sustainability. Examples include the introduction of laser welding in high-strength chassis[2], the use of drones with RFID sensors for material control in trailer yards, and the digitalization of waterless paint booths [3], they illustrate the displacement of physical, repetitive, or hazardous tasks, accompanied by an increase in the demand for technical, digital, and cognitive skills. This pattern aligns with the SBTC approach, as it demonstrates a restructuring of work: automatable tasks tend to disappear or diminish, while those that complement technology become more relevant.

TMMGT's experience demonstrates how technological bias can be socially moderated when implemented within an organizational framework that prioritizes training, internal mobility, and respect for individuals - as advocated by the Toyota Way. At the Guanajuato plant, this approach has materialized in policies that promote training through the dojo system, job rotation, and horizontal work structures that foster continuous learning.

This environment has not only eased adaptation to technological change but has also supported the hiring of women in a traditionally male-dominated industry. While barriers still exist—particularly in leadership roles and physically demanding operations—collected testimonies show their active involvement in key projects. In the pilot design area, for instance, female engineers lead critical vehicle components, coordinate with other plants, and assume high-level technical responsibilities. As Daniel Enríquez stated: "...it is not symbolic; they are responsible for technical compliance, coordination with other plants, and necessary real-time adjustments" (Interview, September 1, 2022).

This case suggests that skill-biased technological change is not an inevitable outcome, but rather a dimension of technological transformation that can be institutionally managed. In this sense, Toyota Guanajuato embodies a model of inclusive modernization, where technology does not replace human labor, but transforms and revalues it through organizational design and investment in workforce capabilities.

This case suggests that the technological bias toward skills is not inevitable—it can be managed through organizational strategies. TMMGT represents a model where technology transforms—but does not displace—human labor, promoting an inclusive and sustainable transition.

4.4 Work Organization: Worker Perceptions and Continuous Training

At TMMGT, production workers understand the importance of technological improvement for maintaining global competitiveness and securing new projects. The introduction of new technologies has been oriented toward enhancing quality, efficiency, and return on investment. Both technical and unionized personnel perceive these changes positively. According to a union representative, automation has not led to staff cuts but rather to a balanced integration between technology and the workforce, reaffirming the strategic value of human labor in the production process.

The flow of information between the company, the union, and the workforce—reinforced by management's open-door policy—has strengthened communication channels and mutual understanding. Within a highly competitive environment, not only among Asian, European, and American assembly plants but also across Toyota's own global subsidiaries, this dynamic has fostered a shared awareness of the need to remain highly competitive in order to continue attracting investment to the plant.

Interview data indicate that automation at TMMGT has not resulted in layoffs. Instead, it has promoted worker upskilling and the strengthening of internal labor markets within the company. This includes job rotation processes that enable employees to acquire new skills and increase their promotion opportunities.

"We assume that if you do your job well in one area, you can also do it in another... with the tools we have in the company, it's all about solving problems."

— Human Resources Manager, TMMGT

The emphasis on soft skills—such as leadership, teamwork, and problem-solving—reinforces the logic of the TPS, in which human capital development is an integral part of the production model.

"I joined Toyota 19 years ago as a maintenance engineer and have rotated through various roles... today I'm a group manager in assembly."

— Group Manager, TMMGT

This talent management model enables technological change to align with upward career trajectories. In this sense, the SBTC approach is complemented by an organizational perspective that not only segments employment but also transforms it from within, prioritizing mobility, training, and stability.

Another important finding is that changes in working conditions—such as increased work pace or the introduction of new shifts—are attributed more to the plant's operational expansion than to the technologies themselves. This positive perception is even shared by the union:

"It's not all robotics; the workforce remains a key factor... It's been a balanced adjustment."

— Union Representative, TMMGT

The TMMGT case thus represents an experience in which the adoption of I4.0 technologies has been carried out without labor conflict, maintaining employment and reinforcing internal training as a tool for technological integration.

To synthesize the empirical findings that link automation patterns with their employment effects and organizational responses, Table 3 summarizes the main results for 2021–2024.

Table 3. Patterns of automation, employment impacts, and organizational strategies at TMMGT (2021–2024)

Automation and digitalization patterns	Observed impacts on employment	Associated organizational strategies
Increase in robots from 127 (2021) to 521 (2024), concentrated in welding, painting, and stamping	Workforce grew from 2,233 to 2,755 employees; no mass layoffs	Gradual and selective automation focused on repetitive, risky, or high-precision tasks
Fluctuating productivity: peak in 2022, drop in 2023 (mismatch between technology and training), recovery in 2024	Temporary adjustment did not result in job loss; reinforced need for continuous training	Dojo-based training and job rotation to align skills with new technologies
Transversal digitalization: drones with RFID, waterless paint booths, real-time traceability	Reconfiguration of job profiles toward cognitive and soft skills (problem-solving, teamwork)	Recruitment of young engineers, collaboration with local universities
Inclusion of women in technical areas and leadership of innovation projects	Diversification of human capital in a traditionally male-dominated sector	Ergonomics and task redesign to enable female participation
Expanding U.S. market context (2021–2024)	Enabled absorption of technologically displaced roles through redeployment and additional hiring	Company-union coordination with open communication and transparent negotiation

Source: Authors' elaboration based on fieldwork from TMMGT (2021–2024).

5. DISCUSSION

Contrary to deterministic projections in much of the automation literature, the integration of advanced technologies at Toyota Guanajuato has not triggered significant workforce displacement. Instead, a combination of organizational design, institutional arrangements, and market conditions, particularly rising U.S. vehicle demand, has enabled automation, robotization, and digitalization to be absorbed without large-scale redundancies or disruptive changes in work organization.

The Skill-Biased Technological Change framework posits that new technologies tend to benefit highly skilled workers while displacing those with lower qualifications, often leading to labor market polarization (Autor, 2015; Acemoglu & Restrepo, 2019). While elements of this pattern are observable at TMMGT, the empirical evidence points to significant nuances that warrant a reinterpretation of SBTC considering adaptive organizational strategies:

a) Automation Without Displacement: Organizational Absorption

At TMMGT, automation has been integrated through redeployment and retraining rather than job cuts. Workers affected by technological changes are reassigned to other areas or upskilled to retain their positions. Even in 2022, a year marked by substantial technological expansion and the introduction of new production lines, there were no mass layoffs. This capacity to "absorb" technological change is consistent with the *reinstatement effect* described by Acemoglu and Restrepo (2019), whereby technology displaces certain tasks but simultaneously generates new ones within the same organization. The combination of internal mobility, targeted training, and gradual technological introduction has allowed TMMGT to maintain employment stability while upgrading operational capabilities.

An element that deserves greater emphasis is the context of U.S. market expansion between 2021 and 2024. The rise in vehicle demand in the United States coincided with the increase in production and employment at TMMGT, which made it easier to absorb potentially displaced positions through internal redeployment and training. This growth environment constitutes a key enabling condition for the "automation without displacement" observed at the plant. However, it is important to reflect on the sustainability of this strategy in scenarios of market stagnation or decline, where the ability to maintain an inclusive model would depend more heavily on organizational decisions and the strength of institutional frameworks.

b) Career Paths: Hiring, Training, and Internal Mobility

Although the hiring of workers with higher educational levels has been favored—particularly through initiatives such as the Fresh Engineers program and the search for young, specialized talent—promotion and professional development processes within the plant are not determined solely by academic background. Interviews with middle managers and union representatives indicate that performance, conduct, and development potential are equally important criteria. Promotions are open to all personnel, provided they meet the organization's established requirements.

> *"[...] in terms of development, promotions, and so on, yes, there have been cases in which we have promoted team members, obviously to team leader positions. Some have already been promoted to group leader, which is a supervisory level, and in some cases to engineers... It is a relatively young plant; people still lack extensive experience; we must consider whether they have completed their degree or not; that's one factor. The other factor is performance—how they have demonstrated their abilities and their potential for development on the job." (Group Manager, Assembly Area, TMMGT)*

"[...] when colleagues are promoted to a higher position—for example, from production staff to team leader—a call for applications is issued with a series of requirements. Obviously, there are also certain aspects that limit who can apply for such a promotion. They cannot have any absences or corrective actions, but the call is open to all workers. We do not directly participate in the promotions; this is handled by the company, but we do encourage all our colleagues to apply and to maintain good conduct, as this is the basis for promotion."
(Union Representative, TMMGT)

The case of TMMGT allows for a reinterpretation of the SBTC approach from an evolutionary perspective. While automation tends to favor more highly skilled workers, in this context qualifications do not function as an exclusionary filter but rather as a potential to be developed. The continuous training strategy—combining technical instruction with soft skills such as leadership, adaptability, and problem-solving—enables initially unskilled workers to transition into specialized technical roles. This suggests that technological bias is not an inevitable outcome, but a phenomenon that can be shaped by organizational structures that promote learning and internal mobility.

c) Diversification of Human Capital and Gender Inclusion

The inclusion of women in technical and engineering roles, supported by ergonomic adjustments and task redesign, has expanded the definition of "skilled" work. These changes illustrate how technology can act as an enabler of diversity when implemented alongside deliberate inclusion strategies (Østergaard *et al.*, 2011; Díaz-García *et al.*, 2013). At TMMGT, female engineers and technicians lead projects in design, quality assurance, and process innovation, challenging historical gender imbalances in the automotive sector. This integration of gender diversity into technological transformation contributes to a broader skills base and enhances organizational innovation capacity.

d) Wage Structure and Polarization Containment

Despite an increase in skill requirements and technological sophistication, wage dispersion remains low (2.2:1 ratio). This relative stability may be attributed to collective bargaining frameworks, retention policies, and the organizational culture embedded in the Toyota Way. Such wage compression can function as a mechanism for preserving workforce cohesion and mitigating internal segmentation (Freeman & Medoff, 1984).

e) More Robots, More Productivity, Less Employment?

The number of robots more than doubled between 2021 and 2024, but productivity did not grow proportionally. This suggests that the success of an automation strategy does not depend solely on the technology installed, but also on the organizational capacity to integrate it, redesign processes, and develop skills.

Taken together, the findings from the Toyota Guanajuato case allow for a more nuanced interpretation of the SBTC approach. Technological expansion has not led to job elimination, but rather to a reconfiguration of work around new skills and roles. The increase in the number of robots has been accompanied by more employment, greater training, and sustained internal mobility, in line with Toyota's ongoing organizational strategy. This positive relationship between technology and labor has been mediated by an organizational culture focused on continuous improvement, a participatory management model, and a production dynamic influenced by international demand—particularly from the U.S. market.

The country of origin and the imported management model also influence how automation is implemented and how employment is protected—or transformed. TMMGT not only adopts and develops advanced technologies but also adapts them to its work environment and aligns them with human development strategies. One example is the pending patents promoted by local engineers to resolve or improve aspects of the production process.

In summary, the case of Toyota Guanajuato demonstrates that automation, robotization, and digitalization do not necessarily imply labor displacement. When accompanied by training, internal mobility, inclusion, and a culture of development, technology can complement—rather than replace—the role of human labor.

f) Labor Agency as a Mediating Factor

The TMMGT case highlights the role of labor agency in shaping the trajectory of automation and digitalization. Beyond managerial strategies, workers and their union actively contribute to the integration of new technologies through continuous training, internal mobility, and transparent union–management dialogue. These dynamics transform the workforce from a passive recipient into a co-manager of technological change. As seen in the development of local innovations and the promotion of women in technical roles, labor agency strengthens organizational resilience and expands the conditions under which technology complements rather than replaces human labor.

g) Reinterpreting SBTC in a Peripheral High-Tech Plant

Taken together, these findings indicate that SBTC is not an inevitable trajectory but rather a contingent process shaped by organisational design, institutional mediation, and strategic choice. At TMMGT, technological change has led to a reconfiguration rather than an elimination of jobs, supported by continuous training, structured mobility, and an inclusive approach to workforce development. The combination of horizontal organisational structures, collaborative union–management relations, and a culture of continuous improvement has moderated potential polarisation effects.

The case also underscores the role of country of origin effects and transferred management models in shaping local automation outcomes. As a subsidiary of a Japanese multinational, TMMGT adopts advanced technologies but adapts them to the local operational environment, embedding them in a people-centred organisational philosophy. Examples include patent applications by local engineers and the in-house development of process innovations, which not only enhance operational efficiency but also strengthen the plant's internal knowledge base.

h) Replicability

The experience of Toyota Guanajuato provides valuable insights into how automation and digitalization can be integrated without triggering labor displacement. Evidence from Ford's Hermosillo plant suggests that certain dynamics—particularly the complementarity between technological upgrading and workforce retraining—may indeed be replicable in other Mexican automotive plants (Carrillo *et al.*, 2023). Moreover, the structural role of the sector, which concentrates almost 70% of the robots installed in Mexico, reinforces the potential for similar trajectories in contexts where technological expansion coincides with organizational adaptation.

Nevertheless, replication is not automatic. The distinctive features of the Toyota Production System—continuous training through the dojo, job rotation, wage compression, and a participatory organizational culture—constitute enabling conditions that are not uniformly present across the industry. Country-of-origin effects and local labor relations significantly shape outcomes: while Japanese management models emphasize gradual, inclusive transformation, other OEMs have implemented automation under different logics, sometimes generating precarization or conflict. Finally, a necessary condition for sustaining the 'virtuous circle' of more technology without job loss is an expanding market, particularly in the United States. In plants where demand does not grow or where institutional frameworks are weaker, replication may be partial or limited.

CONCLUSION

The evidence from TMMGT challenges deterministic narratives that equate automation, robotization, and digitalization with inevitable job losses. In this case, technological integration has been accompanied by deliberate organizational strategies—reassignment, retraining, and job redesign—that have preserved employment while upgrading skills.

The findings indicate that major organizational changes do not coincide with the isolated adoption of automated technologies, but rather with broader transformations, such as the introduction of new vehicle models or the replacement of entire production platforms. In such contexts, companies tend to leverage physical expansion or the opening of new plants to incorporate advanced technologies and best practices. One example is Toyota Guanajuato, which began operations in 2019 with infrastructure based on state-of-the-art technologies.

In this context, automation has been driven not only by the pursuit of efficiency and quality, but also by factors such as occupational safety and environmental sustainability. These technologies have served here as enablers of positive changes in the employment structure. A clear example is the inclusion of women; the introduction of tools and systems that reduce physical effort has enabled greater female participation in production, within an environment historically dominated by men.

A revealing case is that of TMMGT, where these technologies have supported the transformation of skills without triggering layoffs. As documented, workers are reassigned to new areas or functions or trained to retain their positions. This organizational strategy has allowed for the integration of cutting-edge technologies without generating collective resistance or labor conflicts.

No direct link was found between automation and the deterioration of working conditions either. Although technological changes may modify the pace or content of work, these effects appear to be more related to operational expansion than to automation itself. From a wage perspective, collective contracts at TMMGT are revised every two years, and no negative impact on income related to automation or digitalization has been recorded. This finding contrasts with much of the literature on labor polarization and deskilling, which suggests that technological change tends to exclusively benefit highly skilled workers, while displacing or undermining the rest (Autor, 2015; Acemoglu & Restrepo, 2019). According to the Skill-Biased Technical Change hypothesis, one would expect wage concentration at the top of the distribution and erosion of routine work, typically associated with operational tasks. Krzywdzinski (2017) also argues that robotization reduces the qualitative content of work, limiting task autonomy and complexity. However, the data collected at TMMGT not only fail to reflect these effects but instead point to a strategy of productive inclusion through continuous learning, which contrasts with the expected trends of SBTC.

Retraining at the plant is not limited to high-level technical personnel, but also includes operators, suggesting a less exclusionary implementation of technology. This type of empirical evidence nuances the predictions of the skill-biased technical change theory and shows how the institutional context —in this case, the coordination between company, union, and workers— can play a decisive role in shaping the effects of technological change.

The findings of this study suggest that the SBTC theory is only partially applicable to the TMMGT case. Indeed, automation has promoted skills transformation and the reconfiguration of job profiles, in line with the SBTC framework. However, no systematic exclusion of less qualified workers or widespread processes of deskilling or polarization were observed. On the contrary, the company's organizational strategy has broadly promoted retraining and internal mobility, including for operational personnel. This empirical evidence suggests that the effects of technological change on employment depend less on the technology itself and more on how organizations manage its implementation, redesign processes, and develop workforce capabilities. Thus, the TMMGT case offers a more nuanced reading of SBTC, showing that it is possible to integrate advanced technologies without necessarily reproducing their most exclusionary effects.

From a public policy perspective, the TMMGT experience shows that the adoption of Industry 4.0 technologies does not necessarily generate unemployment, as long as it is accompanied by continuous training, internal mobility, and collective bargaining. This points to the need for labor and education policies aimed at strengthening technical competencies and soft skills, as well as institutional frameworks that foster company-union cooperation. Such guidelines would enable the replication of organizational models in which technological innovation translates into employment stability and inclusion, underscoring that the impacts of digitalization depend less on technology itself and more on the organizational and institutional decisions that mediate it.

In sum, the experience of TMMGT demonstrates that technological upgrading in emerging economies is not solely determined by investment and corporate strategy. The active participation of workers and unions—expressed through negotiation, training, and innovation—emerges as a decisive factor in shaping inclusive outcomes. Recognizing labor agency as a central dimension allows us to understand Industry 4.0 not only as a technological process, but also as a social one, where the capacity of workers to influence change is essential to the sustainability of employment.

Contrary to the literature predicting deskilling processes associated with automation, at TMMGT the new tasks derived from Industry 4.0 are cognitively more complex and require greater autonomy than the routines they replaced. Activities such as problem-solving, teamwork, and continuous improvement not only broaden the qualitative content of work but also revalue human capabilities within the organizational model. This evidence strengthens the argument that the integration of advanced technologies does not necessarily entail deskilling, but can instead foster more inclusive trajectories of learning and professionalization.

Future research should further examine how change management models, labour relations, and local institutional environments interact to shape the employment outcomes of Industry 4.0. The TMMGT experience demonstrates that automation's trajectory is not pre-determined: under the right organizational conditions, it can serve as a catalyst for skill upgrading, diversity, and sustainable employment.

REFERENCES

Acemoglu, D. (2002). Technical change, inequality, and the labor market. *Journal of Economic Literature*, *40*(1), 7–72. DOI: 10.1257/jel.40.1.7

Acemoglu, D., & Restrepo, P. (2019). Automation and new tasks: How technology displaces and reinstates labor. *The Journal of Economic Perspectives*, *33*(2), 3–30. DOI: 10.1257/jep.33.2.3

Álvarez, L., & González Marín, M. L. (Coords.). (2017). *Reestructuración productiva de la industria automotriz en la región del TLCAN, 2008–2015*. UNAM.

Andreoni, A., & Anzolin, G. (2025). GVCs meet robots: Smiling, smirking or flattening? The case of the automotive value chain. *. *Innovation and Development*, 1–23. Advance online publication. DOI: 10.1080/2157930X.2025.2468060

Anzolin, G. (2021). *Automation and labour in the age of Industry 4.0: The case of the automotive industry* (ICDD Working Paper).

Arteaga, A. (2020). El patrón de reproducción de capital exportador de especialización productiva y la industria automotriz en México. In Pinto, S., Guevara, A., & Arteaga, A. (Eds.), *La industria automotriz en América Latina: Estudios de las relaciones entre trabajo, tecnología y desarrollo socioeconómico** (pp. 129–160). Universidade Tecnológica Federal do Paraná.

Asociación Mexicana de la Industria Automotriz (AMIA). (2022). *El futuro de la electromovilidad en México*. AMIA.

Autor, D. H. (2015). Why are there still so many jobs? The history and future of workplace automation. *The Journal of Economic Perspectives*, *29*(3), 3–30. DOI: 10.1257/jep.29.3.3

Autor, D. H., Levy, F., & Murnane, R. J. (2003). The skill content of recent technological change: An empirical exploration. *The Quarterly Journal of Economics*, *118*(4), 1279–1333. DOI: 10.1162/003355303322552801

Bensusán, G., & Gómez, W. (2017). Volkswagen de México: Un caso de articulación positiva y fuerte pero inestable entre innovación productiva y social. In J. Carrillo, G. Bensusán, & J. Micheli (Eds.), *¿Es posible innovar y mejorar laboralmente? Estudio de trayectorias de empresas multinacionales en México* (pp. 183–238). UAM Azcapotzalco.

Berman, E., Bound, J., & Griliches, Z. (1994). Changes in the demand for skilled labor within U.S. manufacturing: Evidence from the annual survey of manufactures. *The Quarterly Journal of Economics*, *109*(2), 367–397. DOI: 10.2307/2118467

Bohnsack, R., Pinkse, J., & Kolk, A. (2014). Business models for sustainable technologies: Exploring business model evolution in the case of electric vehicles. *Research Policy, 43*(2), 284–300. DOI: 10.1016/j.respol.2013.10.014

Boyer, R., & Freyssenet, M. (2003). *Los modelos productivos*. Editorial Fundamentos.

Carrillo, J., Martínez, A., López, O., & Díaz, M. (2023). The automotive sector in Mexico: The impact of automation and digitalization on employment. *Research Paper No. 15*. International Labour Organization.

Covarrubias, A. (2020). The boom of the Mexican automotive industry: From NAFTA to USMCA. In Covarrubias, A., & Ramírez, S. (Eds.), *New Frontiers of the Automobile Industry. Exploring Geographies, Technology, and Institutional Challenges** (pp. 323–348). Palgrave Macmillan. DOI: 10.1007/978-3-030-18881-8_13

Covarrubias, A., & Bouzas, A. (2016). *Empleo y políticas sindicales en la industria automotriz de México*. Friedrich Ebert Stiftung.

De Buen, C. (2011). *Los contratos colectivos de trabajo de protección patronal en México. Cuadernos de Análisis Político*. Friedrich Ebert Stiftung.

Díaz-García, C., González-Moreno, Á., & Sáez-Martínez, F. J. (2013). Gender diversity within R&D teams: Its impact on radicalness of innovation. *Innovation (North Sydney, N.S.W.)*, *15*(2), 149–160. DOI: 10.5172/impp.2013.15.2.149

Eurofound. (2018). *Automation, digitisation and platforms: Implications for work and employment*. Publications Office of the European Union.

Frank, A. G., Dalenogare, L. S., & Ayala, N. F. (2019). Industry 4.0 technologies: Implementation patterns in manufacturing companies. *International Journal of Production Economics*, *210*, 15–26. DOI: 10.1016/j.ijpe.2019.01.004

Freeman, R. B., & Medoff, J. L. (1984). *What do unions do?* Basic Books.

García, H., Carrillo, J., & Bensusán, G. (Eds.). (2021). *Salarios en tiempos de libre comercio: ¿Ofrece la industria automotriz salarios dignos en México?* El Colegio de la Frontera Norte.

Goldin, C., & Katz, L. F. (2008). *The race between education and technology*. Harvard University Press., DOI: 10.2307/j.ctvjf9vz

Goos, M., & Manning, A. (2007). Lousy and lovely jobs: The rising polarization of work in Britain. *The Review of Economics and Statistics*, *8*(1), 118–133. DOI: 10.1162/rest.89.1.118

Hirsch-Kreinsen, H., & Hompel, M. (2017). Digitalisierung industrieller Arbeit: Entwicklungsperspektiven und Gestaltungsansätz.

Humphrey, J., & Memedovic, O. (2003). *The global automotive industry value chain: What prospects for upgrading by developing countries?* United Nations Industrial Development Organization (UNIDO). https://www.unido.org/resources/publications/cross-cutting-services/industrial-policy/global-automotive-industry-value-chain

IFR. (2024). Récord de 4 millones de robots en fábricas de todo el mundo. *Sala de prensa IFR*. https://ifr.org/ifr-press-releases/news/record-of-4-million-robots-working-in-factories-worldwide

Kagermann, H., Wahlster, W., & Helbig, J. (2013). *Recommendations for implementing the strategic initiative INDUSTRIE 4.0: Securing the future of German manufacturing industry* (Final report of the Industrie 4.0 Working Group). acatech – National Academy of Science and Engineering. https://www.acatech.de/wp-content/uploads/2018/03/Final_report__Industrie_4.0_accessible.pdf

Kamaruddin, N., Osman, M. R., & Subari, K. (2013). Impact of robotics on employment and motivation of employees in the manufacturing sector. *International Journal of Engineering Research and Applications, 3*(2), 658–665.

Katz, L. F., & Autor, D. H. (1999). Changes in the wage structure and earnings inequality. In O. Ashenfelter & D. Card (Eds.), *Handbook of Labor Economics* (Vol. 3A, pp. 1463–1555). Elsevier. DOI: 10.1016/S1573-4463(99)03007-2

Katz, L. F., & Murphy, K. M. (1992). Changes in relative wages, 1963–1987: Supply and demand factors. *The Quarterly Journal of Economics*, *107*(1), 35–78. DOI: 10.2307/2118323

Keynes, J. M. (1930). Economic possibilities for our grandchildren. In *Essays in Persuasion* (pp. 358–373). W. W. Norton.

Krzywdzinski, M. (2017). Automation, skill requirements and labour-use strategies: High-wage and low-wage approaches to high-tech manufacturing in the automotive industry. *New Technology, Work and Employment, 32*(3), 247–267.

Lampón, J. F., Cabanelas-Lorenzo, P., & Delgado Guzmán, J. A. (2018). Claves en la evolución de México dentro de la cadena de valor global de la industria de autopartes: El caso del Bajío. *El Trimestre Economico*, *85*(339), 483–514. DOI: 10.20430/ete.v85i339.259

Méndez, M. (2025, January 8). La industria automotriz en México: Plantas armadoras de vehículos y su distribución estratégica. *Mexico Industry*. https://mexicoindustry.com/noticia/la-industria-automotriz-en-mexico-plantas-armadoras-de-vehiculos-y-su-distribucion-estrategica

Middlebrook, K. (1988). Industrial restructuring and union politics in the Mexican automobile industry. Paper presented at the conference *México y la Cuenca del Pacífico*.

Morales, J. J., & López, M. G. (2018). Honda y su mercado de trabajo en la planta de El Salto, Jalisco. In de la Garza, E., & Hernández, R. (Eds.), *Configuraciones productivas y laborales en la tercera generación de la industria automotriz terminal en México** (pp. 131–157). UAM / Miguel Ángel Porrúa.

Moreno-Brid, J. C., Gómez Tovar, R., Sánchez Gómez, J., & Gómez Rodríguez, L. (2021). Trade agreements and decent work in Mexico: The case of the automobile and textile industries. *ILO Working Paper No. 36*. International Labour Organization.

OECD. (2020). *The digitalisation of science, technology and innovation: Key developments and policies*. https://doi.org/DOI: 10.1787/b9e4a2c7-en

OICA. (2022). 2021 *Production statistics. International Organization of Motor Vehicle Manufacturers*. http://www.oica.net/category/production-statistics/2019-statistics/

Omar, N. H., Mohd, R. H., & Syaima, N. N. (2025). The impact of automation adoption on employment structures: A longitudinal panel analysis of skill-biased technological change in Malaysia's manufacturing sector, 2012–2023. *International Journal of Academic Research in Progressive Education and Development, 14*(2), 417–431. DOI: 10.6007/IJARPED/v14-i2/25260

Østergaard, C. R., Timmermans, B., & Kristinsson, K. (2011). Does a different view create something new? The effect of employee diversity on innovation. *Research Policy*, *40*(3), 500–509. DOI: 10.1016/j.respol.2010.11.004

Reyes, J., Sánchez, G., & Martínez, M. E. (2018). La instalación de Audi en México: La precariedad laboral como fundamento. In de la Garza, E., & Hernández, R. (Eds.), *Configuraciones productivas y laborales en la tercera generación de la industria automotriz terminal en México** (pp. 107–130). UAM / Miguel Ángel Porrúa.

Richard, R.-B. (2005). Industrialised building systems: Reproduction before automation and robotics. *Automation in Construction*, *14*(4), 442–451. DOI: 10.1016/j.autcon.2004.09.009

Rodríguez, S., & Oropeza, A. (2024, September 2). Toyota revoluciona el futuro de la industria automotriz: Conoce el proceso de fabricación de su planta en Guanajuato. *Mexico Industry*. https://mexicoindustry.com/noticia/toyota-revoluciona-el-futuro-de-la-industria-automotriz-conoce-el-proceso-de-fabricacion-de-su-planta-en-guanajuato

Schumpeter, J. A. (1942). *Capitalism, socialism and democracy*. Harper & Brothers.

Schwab, K. (2016). *The Fourth Industrial Revolution*. Geneva: World Economic Forum._https://law.unimelb.edu.au/__data/assets/pdf_file/0005/3385454/Schwab-The_Fourth_Industrial_Revolution_Klaus_S.pdf

Smit, J., Kreutzer, S., Moeller, C., & Carlberg, M. (2016). *Industry 4.0*. Directorate-General for Internal Policies, European Parliament. https://op.europa.eu/en/publication-detail/-/publication/1b970736-9acb-11e6-868c-01aa75ed71a1

Sturgeon, T. J., Van Biesebroeck, J., & Gereffi, G. (2008). Value chains, networks and clusters: Reframing the global automotive industry. *Journal of Economic Geography*, *8*(3), 297–321. DOI: 10.1093/jeg/lbn007

Sukhodolov, Y. A. (2019). The notion, essence, and peculiarities of Industry 4.0 as a sphere of industry. In Popkova, E. G. (Ed.), *Warsaw Industry 4.0: Industrial revolution of the 21st century*. Springer. DOI: 10.1007/978-3-319-94310-7_1

United Nations Industrial Development Organization (UNIDO). (2022). *Industrial development report 2022: The future of industrialization in a post-pandemic world*. https://www.unido.org/resources-publications-flagship-publications-industrial-development-report

Womack, J. P., Jones, D. T., & Roos, D. (1990). *The machine that changed the world: The story of lean production*. Rawson Associates.

ENDNOTES

1. With collaboration from Ana Karina Hernández, El Colegio de la Frontera Norte (anakarina@colef.mx)

2 This allows the joining of high-strength steels with greater thicknesses together with thinner steels to reduce the overall weight of the vehicle while enhancing its resistance. Manual welding of these materials proved particularly complex due to the contours of the parts, which led Toyota to install six laser welding

lines specifically designed to work with these components (Rodríguez and Oropeza, 2024).

3 In the painting area, new technologies have been incorporated for surface treatment and environmental control. The booths use a dry vapor capture system, without water or the generation of hazardous waste, which improves efficiency and reduces emissions (Paint Group Manager, TMMGT).

Chapter 7
Policy Divergence in China's EV Market:
The Cases of Hangzhou and Tianjin

Haomin Fu
https://orcid.org/0009-0003-6658-669X
Kyushu University, Japan

ABSTRACT

This chapter investigates the significant divergence in EV market outcomes among Chinese cities, despite operating under a uniform national policy framework. Through a comparative case study of Hangzhou and Tianjin, the authors explain why two cities with similar socioeconomic profiles exhibit starkly different market penetration rates. The analysis employs an adapted policy mix framework, deconstructing each city's strategy into three key themes: demand-side regulation, supply-side industrial policy, and infrastructure governance. The findings reveal that Hangzhou's success is attributable to a coherent and consistent policy portfolio where stringent non-fiscal regulations create strong consumer demand. In contrast, Tianjin's slower adoption is linked to a less consistent policy mix constrained by path dependency on legacy automotive joint ventures. The chapter concludes that local state capacity to design and implement a synergistic policy mix is a critical determinant of success in sustainability transitions.

DOI: 10.4018/979-8-3373-4495-9.ch007

1 INTRODUCTION AND BACKGROUND

1.1 The Global Socio-Technical Transition in the Automotive Industry

The global automotive sector is undergoing its most profound transformation in over a century. The shift from internal combustion engine (ICE) vehicles to electric vehicles (EVs) is far more than a simple technological substitution; it represents a complex and far-reaching socio-technical transition that is reconfiguring entire industries, global supply chains, daily user practices, and even geopolitical alignments (Geels, 2019). This transition is not a smooth, linear progression but a dynamic and often contentious process. It pits the immense inertia of the established ICE-based system—with its locked-in investments, entrenched user habits, and powerful incumbent actors—against the disruptive potential of new electric mobility solutions (Magnusson & Werner, 2023). To comprehend the depth and dynamics of this monumental shift, it is essential to first analyze the transition through a robust theoretical lens. This chapter will employ such a framework to deconstruct the forces driving and impeding this global change, before examining how these dynamics manifest at the local level.

The landscape level comprises slow-changing, exogenous factors that create pressure on the existing system. For the automotive industry, these pressures have become immense. Foremost among them are environmental imperatives, including mounting concerns over climate change and the severe public health impacts of urban air pollution, which are compelling governments worldwide to act. The transportation sector is a primary contributor to greenhouse gas emissions in major economies, making its decarbonization a critical policy objective (Bown, 2023). Simultaneously, the transition has become a central arena for geopolitical and economic competition. Nations are vying for leadership in what they perceive as a foundational industry of the 21st century, setting ambitious targets for EV adoption—such as the United States aiming for 50% and China for 40% of total vehicle sales by 2030—to secure a competitive advantage in green technologies and advanced manufacturing (Barwick et al., 2024).

These landscape pressures are destabilizing the incumbent socio-technical regime, which is the deeply entrenched and stabilized system built around the ICE vehicle. This regime is characterized by powerful "lock-in" mechanisms that create immense inertia and resistance to change (Geels, 2019). These include massive sunk investments in factories, supply chains, and refueling infrastructure; established regulations and standards that favor the incumbent technology; deeply ingrained user habits and cultural norms associated with car ownership; and the formidable political and economic power of incumbent automakers and their allied industries, which have historically resisted disruptive change. The transition, therefore, is not a smooth process but a dynamic struggle between the forces seeking to preserve this stable regime and the innovative niches that challenge it.

EVs and their associated technologies initially emerged in such protected niches, nurtured by early adopters, startups, and targeted government support. As landscape pressures intensify, they create "windows of opportunity" for these niche innovations to break through and challenge the dominant regime. The strategic use of industrial policy by governments is the primary mechanism for widening these windows and accelerating the transition. However, the nature of this industrial policy varies significantly across major economic blocs, creating distinct transition pathways.

The United States has recently re-embraced industrial policy with vigor, most notably through the Inflation Reduction Act (IRA). The US approach is primarily a 'demand-pull' and 'reshoring' strategy, utilizing substantial consumer tax credits and manufacturing subsidies to build a domestic EV and battery supply chain (Barwick et al., 2024). A key feature of the IRA is its stringent local content and sourcing requirements, designed to "friend-shore" supply chains and reduce dependence on China. This strategy explicitly intertwines climate goals with national economic security and manufacturing revival, but its protectionist elements have also created significant trade friction with allies, particularly the European Union.

The European Union, by contrast, has historically favored a 'regulatory push' strategy. It has leveraged the power of its large, unified market to impose stringent CO_2 emissions standards on automakers, effectively forcing them to develop and sell EVs to avoid substantial financial penalties (Geels, 2012). This approach works by compelling incumbent firms to innovate within their existing structures. While the EU also provides subsidies, its primary lever has been regulation, and it has been critical of the heavily subsidy-based, and what it views as discriminatory, nature of the US IRA (Barwick et al., 2024).

China represents a third, distinct model: a comprehensive 'state-led ecosystem creation' strategy. From the early stages, Chinese industrial policy has been holistic, aiming to cultivate a complete domestic EV value chain, from raw material processing to battery manufacturing, vehicle assembly, and recycling (Wang, 2024). This approach is characterized by the simultaneous application of multiple policy levers: supply-side mandates like the New Energy Vehicle (NEV) dual-credit system that forces production; demand-side incentives including subsidies and significant non-fiscal benefits; and massive, state-directed investment in the essential complementary sector of charging infrastructure. This multifaceted and centrally coordinated strategy has successfully created the world's largest and most competitive EV market, demonstrating a different pathway to transition that other nations now seek to either counter or emulate.

These divergent global approaches underscore that there is no single path to electrification. The specific design of industrial policy, shaped by national priorities and political economies, creates fundamentally different competitive environments and transition dynamics. It is within this global context of strategic competition and diverse policy philosophies that China's own internal variations, driven by the actions of powerful local governments, become a critical and illuminating subject of study.

1.2 The Chinese Context

China's ascent to global dominance in the electric vehicle market is a direct result of a concerted, long-term national strategy. The central government has identified the EV industry as a strategic imperative, driven by the goals of enhancing energy security by reducing oil import dependency, mitigating severe urban air pollution, and positioning China as a global leader in a high-tech industry of the future (Mazzocco, 2025). This national ambition has been translated into a powerful and comprehensive policy framework designed to create and shape the market from both the supply and demand sides.

Key national-level instruments have been instrumental in this process. The NEV dual-credit policy, for instance, is a formidable supply-side mandate. It requires automakers to achieve a certain quota of NEV production relative to their total sales or forces them to purchase credits from more successful competitors. This policy effectively guarantees a floor for EV production, compelling even reluctant legacy manufacturers to enter the market (Liu & Xie, 2024). On the demand side, China deployed generous national purchase subsidies for over a decade, which were critical in kick-starting the market by reducing the upfront cost barrier for early adopters.

However, an important aspect of China's recent strategy has been the strategic phase-out of these national subsidies, which were fully eliminated for most vehicles at the end of 2022 (Ministry of Finance, 2021). This move signals the government's intent to transition the industry from a policy-driven to a market-driven stage. This policy shift is important for this study because it increases the reliance on non-fiscal and local-level incentives to maintain market momentum. The withdrawal of national subsidies thus acts as a critical juncture—a national-level 'stress test' that reveals the underlying strength, resilience, and strategic intelligence of a city's local policy architecture.

This highlights the crucial role of sub-national governance. While the central government sets the broad strategic direction, the actual implementation of policy is highly decentralized. China's governance structure allows municipal governments considerable autonomy to interpret and adapt national directives to local conditions, leading to substantial variation in policy emphasis, stringency, and ultimately, effectiveness (Lu & Wang, 2022). This decentralization is the source of the central empirical puzzle motivating this chapter: why do cities with ostensibly similar characteristics exhibit such dramatically different EV market outcomes?

To investigate this question, this chapter focuses on the comparative case of Hangzhou and Tianjin. The selection of these two cities is deliberate and analytically powerful. Both are designated “high restriction” cities that limit the issuance of ICE vehicle license plates; both were among the first wave of national EV pilot cities (MoF & MCT, 2009), beginning their promotion efforts at roughly the same time; and, as of 2022, they share comparable socioeconomic profiles in terms of population and GDP (Hangzhou Statistics Bureau, 2022; Tianjin Statistics Bureau, 2023).

Despite these fundamental similarities, their performance in the EV market presents a stark contrast (see Table 1). In 2022, Hangzhou recorded over 213,000 EV sales with a market penetration rate of nearly 42%, while Tianjin saw just under 106,000 sales and a penetration rate of about 31%. To put these figures in context, Hangzhou's 41.86% penetration rate was dramatically higher than China's national average NEV penetration rate, which stood at 25.6% in 2022 (China Government Website, 2023). Hangzhou's cumulative EV ownership also significantly outpaced Tianjin's, at 520,700 versus 374,600 vehicles. This divergence cannot be adequately explained by national policy, which is uniform for both, or by top-line economic indicators, which are similar. The explanation must lie in the distinct local strategies and the specific design of their municipal policy mixes. By deconstructing these local policy portfolios, this chapter aims to uncover the causal mechanisms that translate urban governance choices into distinct market trajectories.

Table 1. Socioeconomic indicators and EV sales for Hangzhou and Tianjin in 2022 (Currency unit: CNY[1])

Indicator	Hangzhou	Tianjin
Population (10 thousand)	1,327.6	1,363.0
GDP (Billion CNY)	18,753	16,311
EV Sales (10 thousand)	21.32	10.59
EV Penetration Rate (EV sales/all passenger car sales)	41.86%	30.57%
EV ownership (10 thousand)	52.07	37.46

Source: China Passenger Cars Association (CPCA), (Tianjin Statistics Bureau, 2023) and (Hangzhou Statistics Bureau, 2022)

2 LITERATURE REVIEW AND THEORETICAL FRAMEWORK

2.1 From Industrial Policy to Policy Mixes

To dissect the complex mechanisms driving the divergent EV adoption trajectories in Hangzhou and Tianjin, it is necessary to move beyond a simple inventory of individual policies. The effectiveness of government intervention in complex socio-technical transitions arises not from any single instrument, but from the strategic combination and interaction of multiple policies working in concert. This requires a theoretical lens grounded in the rich tradition of industrial policy literature, which has evolved significantly over time.

The classic justification for industrial policy, rooted in mainstream economics, is its ability to correct various "market failures," such as information asymmetries, knowledge spillovers, and coordination failures (Gomes et al., 2023). However, this "market-fixing" rationale has been challenged by a heterodox tradition that views the state's role more expansively. Drawing on the legacy of the "developmental state" literature, which analyzed the transformative success of East Asian economies, scholars have long argued that the state does not merely fix markets but actively shapes and creates them. Authors like Wade (1990) and Evans (1995) demonstrated how a cohesive state bureaucracy, simultaneously autonomous from but embedded within the industrial sector, could effectively steer investment and industrial transformation. Scholars like Mazzucato (2016) have conceptualized the "entrepreneurial state" as a lead risk-taker and market creator, particularly at the technological frontier where uncertainty is highest—a condition that perfectly describes the nascent EV industry.

Building on this tradition, contemporary scholars argue that effective industrial policy relies on "policy packages" or "policy mixes" that simultaneously address multiple, interconnected challenges. As Chang and Andreoni (2016, 2020) contend, successful industrial transformation requires a portfolio of policies that are coherently aligned to guide the economy in the desired direction. This perspective is particularly salient for sustainability transitions, which require overcoming not just market failures but also deep-seated systemic and transformational failures that lock in unsustainable technologies and practices (Kern et al., 2019).

Therefore, this chapter adopts the 'policy mix for sustainability transitions' framework, primarily developed by Rogge, Kern, and their colleagues (2019), as its core analytical structure. This framework provides a structured vocabulary to deconstruct and compare the comprehensive policy portfolios deployed by the two cities. A policy mix is understood as comprising three core building blocks:

1. Policy Goals and Strategy: This represents the overarching vision and objectives that guide a government's interventions. The clarity and ambition of this

strategy—whether it prioritizes environmental targets, industrial leadership, or both—fundamentally shape the subsequent choice of instruments.

2. Policy Instruments: These are the specific, tangible tools deployed to achieve the strategic goals. To analyze these instruments systematically, this chapter integrates the complementary framework proposed by Gomes (2023), which emphasizes two critical dimensions for market creation:
   a) Demand Management: This involves shaping the demand structure—the composition of the consumer base and the fiscal versus regulatory pressures upon it. As Chang and Andreoni (2016) argue, managing demand is a frequently neglected but crucial component of industrial policy, as it can enable dynamic increasing returns and "learning by doing".
   b) Sectoral Complementarities: This refers to the crucial relationship between a core technology and its enabling systems. Reminiscent of Rosenstein-Rodan's (1961) "big push" argument, the viability of EVs is inextricably linked to the existence of a robust and reliable charging infrastructure.
3. Policy Mix Characteristics: These are the evaluative dimensions that determine the overall quality and effectiveness of the policy portfolio. This chapter focuses on three key characteristics:
   a) Consistency: The degree of reinforcement or conflict among different policy instruments. A consistent mix features synergistic instruments that amplify each other's effects (Kern et al., 2018).
   b) Coherence: The logical alignment between the policy instruments and the overarching strategic goals. A coherent mix uses tools that are fit for purpose (Kern et al., 2018).
   c) Credibility: The stability, clarity, and long-term predictability of policy signals, which reduces uncertainty and fosters the confidence needed for long-term investment and behavioral change (Rogge & Reichardt, 2016).

A crucial factor underpinning these characteristics is Policy Coordination. As Andreoni and Chang (2019) highlight, misalignment of policies across different government bodies is a primary cause of industrial policy failure. Effective coordination is thus a prerequisite for achieving consistency and coherence, particularly within China's multi-level governance structure where local implementation is key.

It is crucial to note, however, that applying sustainability transition frameworks—including the policy mix perspective often associated with the Multi-Level Perspective (MLP)—to the Chinese context requires critical adaptation. Frameworks developed primarily in European contexts may under-appreciate the unique 'agency' of sub-national actors within China's multi-level governance system. China's governance is often characterized as a form of "experimentalist governance"(Heilmann, 2018), where the central government sets broad strategic directions while empowering local governments to experiment with policy implementation. This structure is a primary driver of the significant "local variations" (Heilmann & Perry, 2020) observed in policy outcomes. Therefore, this chapter employs the policy mix framework not just to evaluate static consistency or coherence, but to specifically illuminate how local governments act as strategic agents who actively shape—rather than passively implement—their own transition pathways within this experimentalist system.

By employing this integrated framework, the analysis shifts from asking "Which single policy is most effective?" to a more sophisticated question: "How does the interaction, synergy, and overall design of a city's complete policy portfolio shape market outcomes?" To operationalize this for the comparative analysis of Hangzhou and Tianjin, this chapter explicitly adapts the analytical framework proposed by Gomes et al. (2023) While their work compared Shenzhen and Suzhou across multiple EV segments, this chapter focuses specifically on the private passenger vehicle market and applies a modified version of their framework to the distinct case of Hangzhou and Tianjin. The investigation is structured around three core analytical themes of local EV policy, as detailed in Table 2.

Table 2. Analytical Framework for Comparing Local EV Policy Portfolios

Analytical Theme	Core Objective	Key Policy Instruments	Primary Evaluation Criteria
Theme 1: Demand-Side Regulation	To create a compelling consumer value proposition for EVs relative to ICEVs.	ICEV License Plate Restrictions: Lottery odds, auction costs. EV License Plate Incentives: Ease of access, cost. Road Access Restrictions: Peak-hour driving bans for ICEVs. - EV Road Access Privileges: Exemptions from driving bans.	Stringency of ICEV Burden: How difficult/costly is it to own and operate an ICEV? Exclusivity of EV Benefit: Is the advantage for EVs unique and significant? Targeting Precision: Does the policy specifically benefit local EV adoption?
Theme 2: Supply-Side Industrial Policy	To foster a dynamic and competitive local industrial ecosystem capable of supplying attractive EV models.	Support for Local Champions: Subsidies, land use, government-firm coordination. Attraction of New Entrants: Incentives for startups and foreign investment. Management of Incumbent Firms: Policies regarding legacy joint ventures.	Coherence with Electrification Goals: Does industrial policy align with EV promotion goals? Path Dependency: Is the local industry locked into the legacy ICE regime? Ecosystem Vibrancy: Does the policy foster innovation, competition, and diversity in the local market?
Theme 3: Infrastructure Governance	To ensure a reliable, accessible, and user-friendly charging network that builds consumer confidence.	Hardware Deployment: Subsidies for public/private pile construction, EV-to-pile ratios. Operational Management: Regulations on maintenance response times, uptime. User Experience Rules: Enforcement against non-charging vehicle occupation ("ICEing"). Service Innovation: Support for new models (e.g., battery swapping, mobile charging).	Quantitative Sufficiency: Is the network dense enough (e.g., service radius)? Qualitative Reliability: Is the network well-maintained and dependable? User-Centricity: Do policies address key consumer pain points and enhance convenience?

Source: Author's own elaboration, partially adopted by Gomes et al. (2023)

3 COMPARATIVE CASE ANALYSISs: HANGZHOU VS. TIANJIN

3.1 Deconstructing the Policy Mixes: A Comparative Analysis

The divergent EV market outcomes in Hangzhou and Tianjin are the direct result of fundamentally different local policy portfolios. While both cities operate under the same national framework and share similar economic stature, a granular analysis of their policy mixes reveals critical differences in strategic intent, instrument design, and implementation rigor across the three analytical themes of demand-side regulation, supply-side industrial policy, and infrastructure governance. Hangzhou has orchestrated a coherent, synergistic mix that creates a powerful, self-reinforcing dynamic for EV adoption, whereas Tianjin's mix is less consistent and coherent, resulting in weaker market signals and more sluggish growth.

3.1.1 Theme 1: Demand-Side Levers — The Primacy of Non-Fiscal Regulation

The core of the divergence between the two cities lies in how they have structured the non-fiscal regulatory environment. This environment shapes the fundamental choice architecture for consumers by altering the relative convenience and cost of owning an EV versus an ICE vehicle.

3.1.1.1 Hangzhou: Engineering a High Opportunity Cost for ICEV Ownership

Hangzhou's government has strategically engineered a demand-side environment that makes choosing and using an ICE vehicle exceptionally difficult and inconvenient, thereby creating a powerful "push" factor away from the incumbent technology. The primary tool is its license plate allocation system. For a resident to acquire a license plate for a new ICE vehicle, they must enter a lottery or a monthly auction. The odds in the lottery are dauntingly low, with an average winning rate of just 0.52%.This near-impossibility forces most prospective buyers into the auction, which imposes a significant financial burden, with the lowest successful bid price recorded at 16,800 CNY(Hangzhou Passenger Vehicle Management Office, 2022). This combination of extremely low probability and high cost creates a formidable barrier that fundamentally alters the calculus of car ownership. In stark contrast, a buyer of a qualifying EV can obtain a local license plate directly and for free, bypassing this entire arduous and costly process.

This powerful "push" is complemented by an equally strategic "pull" in the form of traffic restrictions. In Hangzhou, vehicles with local ICE license plates are barred from driving in the city's core during peak hours on weekdays. Critically, this restriction also applies to all vehicles with non-local license plates, including EVs. The only category of vehicle exempt from all traffic restrictions is an EV with a local Hangzhou license plate (Hangzhou Commercial Bureau, 2022). This policy design is a masterclass in precision. It creates a unique, tangible, and daily benefit that is exclusive to local EV ownership. It directly incentivizes residents not just to purchase an EV, but to purchase and register it in the city, thereby directly stimulating the local market.

3.1.1.2 Tianjin: A Weaker Regulatory Differential

Tianjin, while also classified as a high-restriction city, employs a significantly less stringent set of demand-side regulations. The barrier to obtaining an ICE license plate is considerably lower. The lottery winning rate is 2.81%, more than five times higher than in Hangzhou, making it a statistically more viable option for residents.[2] The minimum auction price is also a more modest 10,000 CNY . The pressure to abandon the prospect of ICE ownership is therefore substantially weaker; the "pain" of sticking with the old technology is less acute.

Furthermore, Tianjin's road access policy, while seemingly pro-EV, is less strategically targeted. In Tianjin, both local ICEVs and foreign-plated ICEVs are restricted during peak hours. However, all EVs, regardless of whether they have a local Tianjin plate or a foreign plate, are exempt from these restrictions (Tianjin Municipal Finance Bureau, 2016). This broad exemption dilutes the unique value proposition of local EV ownership. It fails to create the exclusive privilege seen in Hangzhou, thus providing a weaker incentive for a Tianjin resident to choose an EV over other mobility options. The cumulative effect is a regulatory environment where the opportunity cost of choosing an ICE vehicle is lower, and the unique benefit of choosing a local EV is less pronounced.

3.1.2 Theme 2: Supply-Side Strategies — Ecosystem Cultivation vs. Path Dependency

The attractiveness of an EV market is contingent not only on consumer incentives but also on the availability of a diverse and competitive range of vehicle models. Local industrial policy plays a critical, yet often underestimated, role in shaping this supply-side dynamic. Here again, the strategic divergence between Hangzhou and Tianjin is stark.

3.1.2.1 Hangzhou: Proactive Cultivation of an EV-Native Ecosystem

Hangzhou has proactively cultivated a vibrant, EV-native industrial ecosystem. Its industrial policy is deeply intertwined with the success of Geely Holding Group, a privately-owned automotive giant headquartered in the city. The municipal government has fostered a supportive environment for its local champion, which has aggressively pursued electrification across its diverse brand portfolio, including Geely, the premium Zeekr brand, and Lynk & Co. This symbiotic relationship, which has included innovative governance approaches like assigning government officials to major firms like Geely to improve communication and policy implementation, ensures that the Hangzhou market is supplied with a constant stream of cutting-edge, desirable EV models that cater to various consumer segments (He, 2022). The presence of a dynamic, innovative, and EV-focused anchor firm creates a competitive local market that directly reinforces the city's demand-side policies by providing consumers with attractive products to purchase.

3.1.2.2 Tianjin: Path Dependency on a Legacy Industrial Base

In contrast, Tianjin's automotive industrial landscape has historically been dominated by large, state-owned joint ventures (JVs) with foreign partners, most notably FAW-Volkswagen and FAW-Toyota (Toyota, 2005). These incumbent firms, long profitable from their established ICE vehicle lines, have been demonstrably slower and more reluctant to pivot to full electrification. For years, their EV offerings were limited or less competitive compared to domestic startups and firms like Geely. This delayed transition is evident in plans from FAW-Volkswagen to only introduce models based on its new, China-specific EV platform (the Compact Main Platform) in 2027.

This industrial structure creates a powerful path dependency that hinders the local EV market. Tianjin's industrial policy appears to have been more conservative, focused on preserving its role as a major production base for these legacy JVs rather than aggressively fostering new, EV-native enterprises (Chu, 2023). The result is a local supply of EVs that has been less dynamic and less aligned with the rapid evolution of consumer tastes in the Chinese market. This mismatch between a stated goal of EV promotion and an industrial reality rooted in the ICE era represents a significant incoherence in Tianjin's policy mix.

3.1.3 Theme 3: Infrastructure Governance — Beyond Counting Piles

A robust charging network is a fundamental prerequisite for widespread EV adoption. However, the sheer number of charging piles is an insufficient metric of success. The quality, reliability, accessibility, and overall user experience of the charging infrastructure are what ultimately build or erode consumer confidence. In this dimension of infrastructure governance, Hangzhou has demonstrated a more proactive and meticulous approach than Tianjin.

3.1.3.1 Hangzhou: A Focus on Operational Quality and User Experience

Hangzhou's superiority begins with better fundamental accessibility. The city boasts a more favorable EV-to-public-pile ratio of 3.1:1, compared to 4.28:1 in Tianjin. It also has a denser network, with a public charging service radius of less than 0.7 square kilometers, compared to Tianjin's 0.9 square kilometers (He, 2022; Zuo, 2022). More importantly, Hangzhou's government has moved beyond simply subsidizing construction to actively managing the network's operational quality. The city has implemented specific and enforceable regulations. These rules explicitly prohibit non-charging vehicles from occupying designated charging spaces—a common source of frustration for EV drivers—and mandate that operators must initiate maintenance within one working day of a reported malfunction (Hangzhou Urban and Rural Construction Committee, 2017). This focus on governance ensures a more reliable and less frustrating user experience. Furthermore, the city is pioneering innovative service models, such as deploying autonomous mobile charging robots in its Yuhang district to serve areas with limited fixed infrastructure, showcasing a forward-thinking and user-centric approach (Hangzhou Government, 2025).

3.1.3.2 Tianjin: Reactive Governance and User Frustration

While Tianjin has also implemented policies to encourage and subsidize the construction of charging facilities, the evidence points to a less rigorous focus on operational governance. The city has experienced documented problems with "zombie" piles—stations that are abandoned or unmaintained—and the persistent issue of charging spaces being occupied by ICE vehicles (Sun, 2023; Wan, 2022). Such problems, which are symptomatic of a nationwide challenge of rapid but poorly managed infrastructure rollout, directly undermine consumer confidence and create practical barriers to EV ownership. While Tianjin has made unique efforts, such as building shared charging piles in residential areas (Tianjin Development and Reform Commission, 2023), its overall governance appears more reactive and less effective at ensuring a seamless user experience compared to Hangzhou's proactive and stringent management. This difference in the perceived reliability of the complementary charging sector further contributes to the divergence in market development.

4 SYNTHESIS, DISCUSSION, AND CONCLUSION

4.1 Synthesis: The Power of the Coherent Policy Portfolio

The detailed comparison of Hangzhou and Tianjin's policy choices reveals a clear and compelling conclusion: the significant divergence in their EV market performance is not an accident of circumstance but the logical outcome of the coherence and consistency of their respective local policy mixes. Conversely, Tianjin's more modest results illustrate how a less coherent mix, characterized by weaker signals and internal inconsistencies, can lead to a vicious cycle of sluggishness.

The robustness of Hangzhou's policy mix is further validated by market performance following the complete phase-out of national subsidies at the end of 2022. This national level "stress test" revealed the underlying resilience of the two markets. Hangzhou's market, supported by its strong and coherent non-fiscal incentives, maintained and even accelerated its robust growth, with EV sales increasing from 213,200 in 2022 to 225,558 in 2023 and 307,508 in 2024. In sharp contrast, Tianjin's market, lacking a strong regulatory push, saw a temporary increase to 154,488 units in 2023 before declining significantly to just 81,900 units in 2024—falling well below its 2022 baseline. While this study does not employ the econometric models necessary to quantify the precise impact of each policy lever, this stark post-subsidy divergence strongly supports this chapter's central thesis: a coherent, synergistic policy portfolio, as seen in Hangzhou, creates a self-sustaining market that is resilient to national policy shocks, whereas Tianjin's less consistent mix proved vulnerable.

Applying the policy mix framework, Hangzhou's approach excels on all key characteristics. The city's policy mix is highly coherent, with its instruments all logically aligned with a clear and ambitious strategic goal: to establish the city as a leader in the EV transition. The stringent demand-side regulations, the proactive cultivation of a local EV industrial champion, and the meticulous governance of charging infrastructure all work in unison towards this single objective. The mix is also highly consistent, as the different policy elements actively reinforce one another. The strong regulatory "push" on consumers creates a large and predictable local market, which in turn de-risks investment for supply-side actors like Geely. The resulting abundance of attractive EV models and a reliable charging network then validates the consumer's decision to switch, creating a powerful positive feedback loop. Finally, the mix is highly credible. The city's unwavering commitment to stringent, long-term policies sends a clear and unambiguous signal to both industry and consumers, fostering the confidence needed for investment and adoption.

Tianjin's policy mix, in contrast, exhibits weaknesses across these same dimensions. It demonstrates lower coherence, as a potential conflict exists between the official goal of promoting EVs and the implicit industrial policy of supporting legacy ICE-focused joint ventures. This creates mixed signals and a lack of clear strategic direction. The mix is also less consistent. The moderate regulatory push is not sufficiently powerful to overcome the inertia of the local industrial base or to compensate for the frustrations of a less reliable charging network. The different policy elements do not amplify each other to the same extent as in Hangzhou. This leads to lower credibility, as the weaker and sometimes conflicting signals create greater uncertainty for consumers and do less to incentivize rapid change from incumbent industrial players.

Hangzhou's success is a textbook example of how a synergistic portfolio of mutually reinforcing policies can create a virtuous cycle of market growth (see Figure 1).

Figure 1. ***The Virtuous Cycle of Hangzhou's Policy Mix***

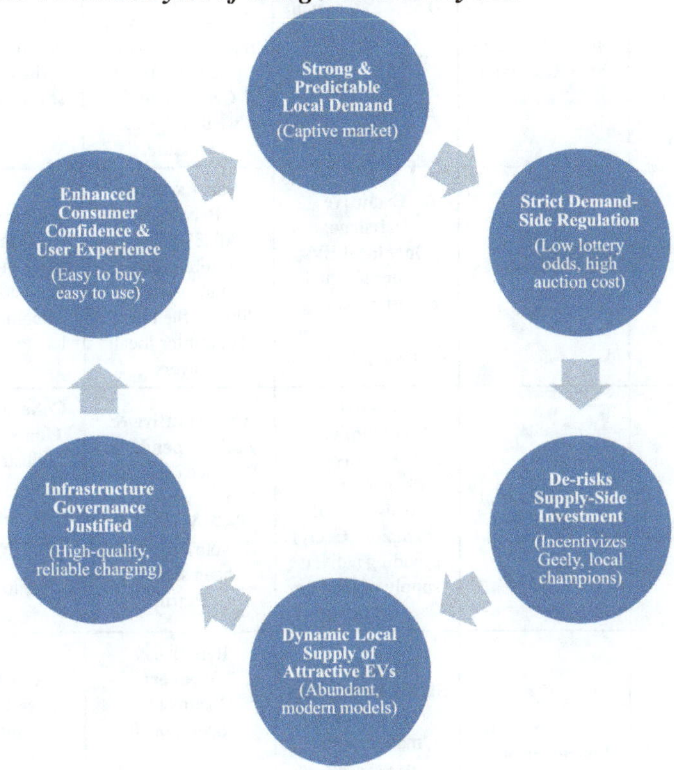

This analysis can be synthesized into a "policy portfolio" model, which posits that the synergistic combination of policies is the primary explanatory variable for the divergent outcomes. This finding aligns with broader academic research on China's EV market, which increasingly emphasizes that the effects of policy mixes—rather than individual instruments—are the key drivers of adoption, and that these effects vary significantly based on local conditions (Zhao et al., 2024). The Table 3 below provides a structured summary of this comparative analysis.

Table 3. Comparative Synthesis of the EV Policy Mix in Hangzhou and Tianjin

Policy Dimension	Policy Element	Hangzhou's Approach	Tianjin's Approach	Evaluation of Mix Characteristic
Demand-Side Regulation	ICEV License Plate Restriction	**Highly Stringent:** 0.52% lottery win rate; high auction price of 16,800 CNY. Creates immense "push" pressure.	**Moderately Stringent:** 2.81% lottery win rate; lower auction price of 10,000 CNY. Creates weaker "push" pressure.	**Stringency:** Hangzhou's policy is substantially more stringent, creating a stronger incentive to switch.
	EV Road Access Priority	**Exclusive Privilege:** Only local EVs are unrestricted, creating a strong, targeted incentive for local purchases.	**Non-Exclusive Privilege:** All EVs (local & foreign) are unrestricted, diluting the unique benefit for local buyers.	**Targeting & Consistency:** Hangzhou's policy is more consistent with the goal of driving local market growth.
Supply-Side Industrial Policy	Local Industrial Structure	**Proactive & Dynamic:** Active cultivation of a private, EV-focused local champion (Geely) providing a diverse supply of modern EVs.	**Conservative & Path-Dependent:** Reliance on legacy state-owned JVs (FAW-VW, FAW-Toyota) that have been slow to electrify.	**Coherence with EV Goals:** Hangzhou's industrial policy is highly coherent with its electrification strategy. Tianjin's exhibits incoherence due to incumbent inertia.
Infrastructure Governance	Management & User Experience	**Proactive & Strict:** Enforceable regulations on maintenance, usage (anti-ICEing), and innovative service models (e.g., mobile charging).	**Reactive & Weaker:** Documented operational issues including abandoned piles and misuse of charging spaces, eroding user confidence.	**Credibility & User Confidence:** Hangzhou's meticulous governance builds higher user confidence and network credibility.
Overall Policy Mix	**SYNTHESIS**	**Coherent, Consistent, Credible:** Mutually reinforcing policies create a virtuous cycle of accelerated adoption.	**Incoherent, Inconsistent, Less Credible:** Weaker, partially conflicting signals create a vicious cycle of sluggish growth.	**Overall Effectiveness:** Hangzhou's synergistic portfolio is demonstrably more effective at creating a dynamic, self-sustaining market.

Source: Author's own elaboration

In essence, Hangzhou's experience demonstrates a virtuous cycle: strong demand-side pressure creates a captive market, which fuels a dynamic local supply of vehicles and justifies robust investment in well-managed infrastructure. This positive experience reinforces consumer choice and accelerates the entire transition. Tianjin's experience points toward a more vicious or stagnant cycle: a weaker regulatory push provides less urgency for consumers, which in turn gives legacy industrial players less incentive to rapidly change, while a less reliable infrastructure network adds another layer of friction. Each element of policy weakness exacerbates the others, leading to a market that fails to achieve its full potential.

4.2 Discussion

The detailed comparative analysis of Hangzhou and Tianjin offers insights that extend far beyond the specific context of these two cities, contributing to broader academic debates on urban governance, industrial policy, and the dynamics of sustainability transitions. The findings provide actionable lessons for policymakers globally who are grappling with the challenge of steering complex, large-scale transformations like the shift to electric mobility.

4.2.1 The Strategic Local State

First, this chapter makes a significant contribution to the literature on urban governance and the role of the local state. It empirically demonstrates that in a large, centrally planned economy like China, the local government is not merely an administrative implementer of national directives but a powerful strategic actor in its own right. The study challenges any monolithic view of "China's EV policy" by revealing how divergent local strategies, capabilities, and industrial histories lead to profoundly different outcomes on the ground. The case of Hangzhou and Tianjin shows that local state capacity—defined here as the ability to design and implement a coherent, consistent, and credible policy mix—is a critical variable in determining the success of national transition goals.

4.2.2 Enriching Transition Theory

Second, the findings reinforce and enrich core concepts from sustainability transitions theory. The case of Tianjin serves as a classic illustration of how incumbent regimes and path dependency can create powerful barriers to change. The city's reliance on its established, ICE-focused joint ventures created an industrial inertia that its moderate policy incentives were insufficient to overcome. Hangzhou's story, conversely, highlights how a transition can be accelerated. The city's government effectively channeled landscape-level pressures (like the national push for electrification) through stringent local regulations, while simultaneously nurturing a powerful local 'niche' actor (Geely) that could challenge and ultimately reshape the local automotive regime. This demonstrates a successful pathway of strategic niche management combined with institutional pressure.

4.2.3 Broader Implications: Equity and Access

Third, this analysis reveals a critical trade-off in Hangzhou's highly effective policy mix: market efficiency versus socio-economic equity. This chapter has focused on adoption metrics, but the reliance on stringent non-fiscal barriers raises significant equity challenges, a topic under-emphasized in the initial analysis.

The Hangzhou model's core mechanism—a highly restrictive lottery and costly auction system (with bids recorded at 16,800 CNY) —is an inherently regressive tool. While it powerfully 'pushes' consumers away from ICEVs, it simultaneously creates a steep mobility barrier for low-income groups. It effectively establishes a choice architecture where only wealthier residents can afford *either* the auction price for an ICEV or the (often higher) upfront cost of a new EV. Tianjin's lower auction barrier (10,000 CNY) is comparatively less regressive but still risks entrenching existing mobility inequalities.

Furthermore, this chapter's focus on aggregate municipal data obscures the critical urban-rural divide in charging access, a key socio-spatial dimension of the transition. The dense charging network celebrated in urban Hangzhou (with a service radius under 0.7 sq km) reflects a policy choice prioritizing high-visibility, high-adoption-potential areas. This logic, however, inevitably creates "charging deserts" in peri-urban and rural districts. This geographic disparity structurally limits EV adoption to affluent, urban-dwelling households and fails to address the mobility needs of the broader population, a key challenge for the next, more inclusive phase of the transition. This highlights a potential blind spot in policy mixes that are optimized purely for market creation rather than for holistic social transformation.

4.2.4 Broader Implications: Environment and Grid

Fourth, the analysis must extend to the foundational assumption of the EV transition: its environmental benefits. While this chapter does not provide a direct quantitative assessment of emission reductions, the divergent market outcomes strongly imply different environmental trajectories. Hangzhou's significantly higher EV penetration rate (41.86% in 2022) suggests a more substantial reduction in localized, street-level air pollution (e.g., NOx and PM2.5) compared to Tianjin (30.57%).

However, this localized benefit must be distinguished from the *net climate benefit* (i.e., total greenhouse gas reduction), which is contingent upon a critical systemic challenge: the carbon intensity of the power grid. If a city's rapid EV adoption is powered primarily by a coal-dominated national electricity grid, the decarbonization benefits are substantially muted, effectively 'shifting' emissions from the vehicle's tailpipe to the power plant. This highlights a potential incoherence in the broader national strategy, where transportation policy (promoting EVs) may be out of sync with energy policy (decarbonizing the grid).

This reveals a crucial lesson for policymakers globally: an effective green transition policy mix cannot focus on vehicle adoption in isolation. It must be systemically and coherently linked with policies that simultaneously accelerate the decarbonization of the energy grid itself.

4.2.5 Global Context: Policy Lessons and Geopolitics

Fifth, the chapter's findings offer internationally applicable lessons for policy design. The central conclusion—that a synergistic and coherent policy portfolio is more effective than a collection of strong but isolated instruments —is a principle of universal relevance. Policymakers in Europe, North America, and other emerging economies can draw several key lessons. Notably, the finding that stringent non-fiscal regulations (which increase the cost and inconvenience of the incumbent technology) can be more powerful and cost-effective than direct fiscal subsidies is particularly salient in an era of constrained public budgets.

A useful global parallel can be drawn with Norway, another world leader in EV adoption. While highly effective, Norway's model has been overwhelmingly reliant on massive fiscal incentives (e.g., full exemptions from VAT and purchase taxes), a costly 'demand-pull' strategy that is difficult for nations with constrained budgets to replicate. Hangzhou's experience, in contrast, provides a powerful and complementary lesson. It demonstrates a pathway with potentially greater cost-effectiveness by prioritizing a stringent non-fiscal 'regulatory push' to shape the consumer choice architecture.

This chapter's domestic focus must be contextualized within the post-2022 global landscape of rising geopolitical tensions. The European Union's anti-subsidy investigation into Chinese EVs, for example, creates new pressures and vulnerabilities for local industrial ecosystems. This external pressure impacts the two cities differently: Hangzhou's local champion, Geely, which has a significant global footprint (e.g., Volvo, Polestar), faces direct export challenges. Conversely, Tianjin's industrial base, reliant on legacy JVs with German and Japanese partners (FAW-VW, FAW-Toyota), is embedded in different global value chains that face distinct, though related, trade and technology-sharing pressures. This illustrates that future local industrial policy must navigate not only domestic competition but also this complex web of international trade and technology "friend-shoring."

4.2.6 Future Trajectories: The CASE Revolution

This analysis provides a crucial insight into the foundations of future mobility. A successful local EV market is not an end, but a foundational layer for the next, far more complex technological wave of CASE mobility. The divergent trajectories of Hangzhou and Tianjin in the EV transition position them differently for this next frontier.

Hangzhou is building more than just a large user base; it is cultivating an entire ecosystem. By fostering a dynamic local industry with deep expertise in electrification and software (led by Geely), and by developing a robust and intelligently managed energy infrastructure, the city is positioning itself as a leader. These elements are prerequisites for deploying autonomous driving (which requires advanced sensors and vehicle platforms), vehicle-to-grid (V2G) services (which require a reliable, well-managed charging network), and integrated smart mobility platforms.

In contrast, cities that lag in the initial EV transition risk being left behind. Tianjin, with its industrial base still grappling with path dependency on legacy JVs and a less reliable charging network, lacks the critical mass of users, the cutting-edge industrial capability, and the advanced infrastructure governance needed to compete in the more advanced and lucrative CASE service economy. This demonstrates that the policy choices made today in the EV transition are path-creating, setting the stage for regional leadership—or marginalization—in the next technological revolution.

4.2.7 Limitations and Future Research

Finally, this chapter's conclusions must be considered in light of its limitations, which in turn suggest avenues for future research.

First, the analysis relies on descriptive data (as shown in Table 1) and secondary sources to build its causal argument for the "policy portfolio model." It does not employ econometric models to quantify the precise cost-benefit of specific policy instruments or their exact contribution to adoption rates. Future research should aim to develop quantitative models that isolate the impact of non-fiscal regulations (like Hangzhou's lottery) versus supply-side incentives (like support for Geely) to validate the synergies proposed here.

Second, the study is based on publicly available data and reports, lacking primary data from stakeholder interviews or surveys. This limits the analysis of internal government coordination and the specific "consumer behavior" driving choices in each city. Future qualitative research, involving interviews with local policymakers and consumer focus groups, would be invaluable in uncovering the deeper political-economy and behavioral mechanisms at play.

Third, as acknowledged in Sections 4.2.3 and 4.2.4, the chapter under-emphasizes the crucial dimensions of socio-economic equity and detailed environmental outcomes. These areas were not the primary focus but represent critical avenues for subsequent research to build a more holistic understanding of the transition's true costs and benefits.

4.3 Conclusion

As the global EV market continues to mature, the direct role of policy support will inevitably diminish. The ultimate goal of any industrial transition policy is to create a market that is vibrant, competitive, and self-sustaining, thus rendering the policy itself obsolete. This study demonstrates that the path taken to reach that mature state matters immensely. The Hangzhou model—characterized by the strategic construction of a coherent policy mix that aligns demand-side pressures, supply-side dynamism, and infrastructure reliability—offers a more effective and accelerated pathway to building such a resilient, self-sustaining market. It serves as an essential reference not only for understanding the nuances of China's remarkable EV transition but also for informing the design of effective transition policies worldwide.

REFERENCES

Andreoni, A., & Chang, H.-J. (2019). The political economy of industrial policy: Structural interdependencies, policy alignment and conflict management. *Structural Change and Economic Dynamics*, *48*, 136–150. DOI: 10.1016/j.strueco.2018.10.007

Barwick, P., Kwon, H., Li, S., Wang, Y., & Zahur, N. B. (2024, December 13). *Industrial policies and innovation in the electrification of the global automobile industry*. CEPR. https://cepr.org/voxeu/columns/industrial-policies-and-innovation-electrification-global-automobile-industry

Bown, C. P. (2023). *Industrial policy for electric vehicle supply chains and the US-EU fight over the Inflation Reduction Act | PIIE* (No. Industrial policy for electric vehicle supply chains and the US-EU fight over the Inflation Reduction). The Peterson Institute for International Economics. https://www.piie.com/publications/working-papers/2023/industrial-policy-electric-vehicle-supply-chains-and-us-eu-fight

Chang, H.-J., & Andreoni, A. (2016). *Industrial Policy in a Changing World: Basic Principles, Neglected Issues and New Challenges*. Cambridge Journal of Economics 40 Years Conference. https://www.semanticscholar.org/paper/Industrial-Policy-in-a-Changing-World%3A-Basic-Issues-Chang-Andreoni/e29246a6311db07a85006d33e4ac5cc6446cb90b?p2df

Chang, H.-J., & Andreoni, A. (2020). Industrial Policy in the 21st Century. *Development and Change*, *51*(2), 324–351. DOI: 10.1111/dech.12570

China Government Website. (2023, January 24). *China's New Energy Vehicle Production and Sales Rank First Globally for Eight Consecutive Years*. https://www.gov.cn/xinwen/2023-01/24/content_5738622.htm

Chu, Y. (2023). *EV city profile: Passenger car market, Tianjin*. ICCT – International Council on Clean Transportation Europe. https://theicct.org/publication/ev-city-profile-passenger-car-market-tianjin/

Evans, P. (1995). *Embedded Autonomy: States and Industrial Transformation*. Princeton University Press., https://www.jstor.org/stable/j.ctt7t0sr DOI: 10.1515/9781400821723

Geels, F. W. (2012). A socio-technical analysis of low-carbon transitions: Introducing the multi-level perspective into transport studies. *Journal of Transport Geography*, *24*(C), 471–482. DOI: 10.1016/j.jtrangeo.2012.01.021

Geels, F. W. (2019). Socio-technical transitions to sustainability: A review of criticisms and elaborations of the Multi-Level Perspective. *Current Opinion in Environmental Sustainability*, *39*, 187–201. DOI: 10.1016/j.cosust.2019.06.009

Gomes, A. D. P., Pauls, R., & ten Brink, T. (2023). Industrial policy and the creation of the electric vehicles market in China: Demand structure, sectoral complementarities and policy coordination. *Cambridge Journal of Economics*, *47*(1), 45–66. DOI: 10.1093/cje/beac056

Hangzhou Commercial Bureau. (2022). *Implementation Rules for Subsidies for Purchase of New Energy Vehicles in Hangzhou, 2022*. https://www.hangzhou.gov.cn/art/2022/6/30/art_812262_59060267.html?eqid=932dc7c90000f0f50000000464891e8e

Hangzhou Government. (2025, February 21). *Autonomous charging vehicles hit Hangzhou city's streets*. https://www.ehangzhou.gov.cn/2025-02/21/c_292710.htm

Hangzhou Passenger Vehicle Management Office. (2022). *Hangzhou license plate lottery winning rate*. http://hz.bendibao.com/live/20201123/96674.shtm

Hangzhou Statistics Bureau. (2022). Hangzhou. *Statistical Yearbook. Annuaire Statistique, 2022*, •••. http://tjj.hangzhou.gov.cn/art/2022/11/28/art_1229453592_4107024.html

Hangzhou Urban and Rural Construction Committee. (2017). *Notice on the Issuance of Interim Measures for the Operation and Management of Public Charging Piles for New Energy Electric Vehicles in Hangzhou* (No. Hangzhou Construction Science and Technology Development [2017] No. 70). http://www.caam.org.cn/chn/9/cate_104/con_5207545.html

He, M. (2022, September 28). *Business Geography of China's NEV Cities⑦|Hangzhou: The First Internet City's "Buddhist" Competition*. https://www.21jingji.com/article/20220928/herald/f656da9b76eb8a547e7e5593b4d563ff.html

Heilmann, S. (2018). *Red Swan: How Unorthodox Policy-Making Facilitated China's Rise*. The Chinese University of Hong Kong Press.

Heilmann, S., & Perry, E. J. (2020). *Mao's Invisible Hand*. Brill. https://brill.com/edcollbook/title/58480

Kern, F., Kivimaa, P., Rogge, K., & Rosenow, J. (2018). Policy mixes for sustainable energy transitions: The case of energy efficiency. In *Transitions in Energy Efficiency and Demand*. Routledge. DOI: 10.4324/9781351127264-12

Kern, F., Rogge, K. S., & Howlett, M. (2019). Policy mixes for sustainability transitions: New approaches and insights through bridging innovation and policy studies. *Research Policy*, *48*(10), 103832. DOI: 10.1016/j.respol.2019.103832

Liu, Z., & Xie, T. (2024). The Impact of Policy Thematic Differences on Industrial Development: An Empirical Study Based on China's Electric Vehicle Industry Policies at the Central and Local Levels. *Energies*, *17*(22), 22. Advance online publication. DOI: 10.3390/en17225805

Lu, S., & Wang, H. (2022). Limited Decentralization: Understand China's Land System from the Perspective of Central-Local Relation. *Land (Basel)*, *11*(4), 4. Advance online publication. DOI: 10.3390/land11040517

Magnusson, T., & Werner, V. (2023). Conceptualisations of incumbent firms in sustainability transitions: Insights from organisation theory and a systematic literature review. *Business Strategy and the Environment*, *32*(2), 903–919. DOI: 10.1002/bse.3081

Mazzocco, I. (2025). *Electric Vehicles, China, and the Industrial Strategies Reshaping Mobility in Emerging Economies*. Center for Strategic and International Studies. https://www.csis.org/analysis/electric-vehicles-china-and-industrial-strategies-reshaping-mobility-emerging-economies

Mazzucato, M. (2016). From market fixing to market-creating: A new framework for innovation policy. *Industry and Innovation*. https://www.tandfonline.com/doi/abs/10.1080/13662716.2016.1146124

Ministry of Finance. (2021). *Notice on the Financial Subsidy Policy for the Popularization and Application of New Energy Vehicles in 2022* (No. Financial Construction [2021] No. 466). https://www.gov.cn/zhengce/zhengceku/2021-12/31/content_5665857.htm

MoF, & MCT. (2009). *Notice on the development of energy saving and NEVs Demonstration and Promotion Pilot Program* (No. Finance [2009] No. 6). https://www.gov.cn/zwgk/2009-02/05/content_1222338.htm

Rogge, K. S., & Reichardt, K. (2016). Policy mixes for sustainability transitions: An extended concept and framework for analysis. *Research Policy*, *45*(8), 1620–1635. DOI: 10.1016/j.respol.2016.04.004

Rosenstein-Rodan, P. N. (1961). Notes on the Theory of the 'Big Push.' In H. S. Ellis (Ed.), *Economic Development for Latin America: Proceedings of a Conference held by the International Economic Association* (pp. 57–81). Palgrave Macmillan UK. DOI: 10.1007/978-1-349-08449-4_3

Sun, N. (2023, July 31). New energy charging problem caused anxiety? Zhejiang program to help this "pile" small matter... ... *China News - Zhejiang*. https://www.zj.chinanews.com.cn/nzxzl/2023-07-31/detail-ihcrqyau6706271.shtml

Tianjin Development and Reform Commission. (2023, March 2). *Municipal Development and Reform Commission promotes charging infrastructure construction Driving consumption of NEVs*. https://fzgg.tj.gov.cn/xxfb/xwxx/202303/t20230302_6126366.html

Tianjin Municipal Finance Bureau. (2016). *Municipal Finance Bureau Municipal Science and Technology Commission on the issuance of Tianjin to promote the application of new energy vehicles local subsidy management interim measures notice* (No. Tianjin Financial Construction I [2016] No. 40). https://cz.tj.gov.cn/zwgk_53713/zcwjx/zcwjst/202007/t20200719_2962749.html

Tianjin Statistics Bureau. (2023). *Tianjin National Economic and Social Development Statistics Bulletin 2022*. https://stats.tj.gov.cn/tjsj_52032/tjgb/202303/t20230317_6142668.html

Toyota. (2005, March 22). *Tianjin FAW Toyota Motor Co., Ltd. Begins Production of the Crown in China | Toyota Motor Corporation Official Global Website*. https://global.toyota/en/detail/251090

Wade, R. (1990). *Governing the Market: Economic Theory and the Role of Government in East Asian industrialization*. Princeton University Press., https://www.researchgate.net/publication/272933957_Governing_the_Market_Economic_Theory_and_the_Role_of_Government_in_East_Asian_Industrialization DOI: 10.1515/9780691187181

Wan, X. (2022, January 20). Tianjin Consumers Association survey shows: New energy vehicle charging pile should also "charge charge." *China Consumers' Daily*. http://auto.ce.cn/auto/gundong/202201/20/t20220120_37272958.shtml

Wang, X. (2024). *The indispensable role of industrial policy: The case of the Chinese EV industry*. https://www.osservatorio-economie-emergenti-torino.it/emerging-economies/78-26-january-2024/443-the-indispensable-role-of-industrial-policy-the-case-of-the-chinese-ev-industry.html

Zhao, X., Li, X., Jiao, D., Mao, Y., Sun, J., & Liu, G. (2024). Policy incentives and electric vehicle adoption in China: From a perspective of policy mixes. *Transportation Research Part A, Policy and Practice*, *190*(C), 104235. https://ideas.repec.org//a/eee/transa/v190y2024ics0965856424002830.html. DOI: 10.1016/j.tra.2024.104235

Zuo, M. (2022, October 28). *Business Geography of China's NEV Cities |Tianjin: An Old Auto Industry City Waiting for a Turnaround*. https://m.21jingji.com/article/20221028/herald/daf738e8ea30ff6087980e3e0d8fe4cf.html

ENDNOTES

1 Average USD to CNY exchange rate in 2022: 1 USD=6.73 CNY

Chapter 8
Evaluating Customer Satisfaction With Electric Vehicle (EV) Service and Quality

Mohammed Nagro
Effat University, Saudi Arabia

Burhan Uluyol
https://orcid.org/0000-0002-9984-489X
Istanbul Sabahattin Zaim University, Turkey

Mahmoud Husain
Effat University, Saudi Arabia

Jamal Thiqah
Effat University, Saudi Arabia

ABSTRACT

Electric vehicles (EVs) have attracted increased attention from researchers and policymakers due to their role in sustainable transportation. This study aims to determine factors influencing drivers' satisfaction with EV service and quality. The research examines four key factors: service quality, vehicle quality, pricing, and customer experience. Using a structured questionnaire, data was collected from 315 respondents who received service from EV providers. Multiple regression analysis revealed that service quality, vehicle quality, and pricing have positive and

DOI: 10.4018/979-8-3373-4495-9.ch008

significant impacts on customer satisfaction, while customer experience showed no statistical significance. This study provides the first comprehensive analysis of EV owners' satisfaction based on their service and usage experiences.

1. INTRODUCTION

Electric vehicles (EVs) have attracted increased attention from researchers and policymakers due to sustainable transportation. While promoting EV adoption is a priority, customer satisfaction remains crucial for long-term success. Dissatisfied customers may undervalue EVs, adversely affecting their sustainable development. In a competitive market, companies must deliver high-quality service, as customer satisfaction directly links to service quality, long-term relationships, and loyalty (Dam and Dam, 2021).

Promotion of EVs and the popularization of these modes of transportation has become an important imperative in reducing air pollution and greenhouse gas emissions. Inadequate service and quality are reasons why many refuse to maintain or purchase an EV. Dissatisfaction with service and quality can lead to a negative assessment of the goods offered by the supplier of such a service. Because electric vehicles are a qualitatively new product, it is important to study the factors that can influence the quality of the EV service at different stages of its life cycle. First and foremost, the quality of the car is extremely important for a potential buyer, ensuring the safety and long life of the vehicle (Goel et al., 2021). Nevertheless, a small number of studies have investigated the factors affecting service quality in EV sales and maintenance.

The development of EVs has gradually been establishing a trend in the transportation field. The expansion of electric vehicles has also allowed car users to alter their driving habits, with electric vehicles transforming from a simple means of transportation into a user experience delivered by service industry domains. How to measure user satisfaction becomes something that electric vehicle operators must emphasize and understand, and it becomes particularly important to correspond to the implementation of a service quality decision strategy (Kwon et al., 2020). This study targets the users of electric vehicles; by means of a structural covariance model, it will verify the service quality experienced by electric vehicle users and the critical influential relationships of satisfaction with electric vehicles.

EVs are important for the improvement of urban environments and the reduction of greenhouse gas emissions. Negative factors that can influence the use of electric vehicles must be eliminated. This research explores the factors that influence user satisfaction during interactions with a fast-charging electric vehicle service and provides greater insight into user attitudes towards the service. User satisfaction is essential for the popularization of electric vehicles; consequently, it is important to know the opinions of users about the fast charge points. This study has discovered that users are satisfied with the quality of the service, the charging time, and the support provided by the operator. However, users require that the fast charge points are clean at all times and are near other services. Users also think that the handling of the cable connection should be improved. The fundamental aspects that led me to carry out this type of investigation are: - It seeks to improve the urban environment by reducing environmental impacts, including the negative externalities produced by transportation. - It seeks to ensure that citizens do not view the use of electric vehicles as a loss of comfort, as this would further delay the desire to change their means of transportation (Su et al., 2020). It can contribute to the technological development of electric vehicles, helping manufacturers and operators understand the opinions of users regarding the quality and technical characteristics of the product or service.

Developing an EV service is important to established companies for several reasons. One motive is the need to globally diversify the energy sources upon which our transportation systems depend. Another motive stems from the traditionally important customer-product interaction aspect of vehicles. To successfully integrate electric vehicles into customer transportation systems, we must better understand those boundary conditions (Aljarboa, 2024). However, little academic research has examined customer service and product quality factors that might be important determinants of customer satisfaction with EVs. The primary objective of this study is to initiate such research. This paper sets forth an initial list of determinants and tests those that are data available on the current Saudi market introduction of EVs. The dependent variables forming the focus of the study are two aspects of customer satisfaction when they are driving an electric vehicle

To date, there is limited research which studied the factors influencing service quality in operation and how unbearable service quality can affect the operation of the EV. As many new players have entered the electric vehicle market, it is important to identify key aspects of customer interaction that can help better understand how to win and maintain customers (Shokouhyar et al., 2020). The paper contributes to the existing literature by identifying the key factors affecting the quality of the after-sales service of electric vehicles and showing the impact of low-quality service on the main characteristics of electric vehicles and their safety in use.

To better understand consumers and offer more efficient services, companies increasingly analyse relationships between different satisfaction dimensions and respond to the change in consumer needs to more accurately define levels or factors of customer satisfaction with a service. For theoretically sound research, it is important to identify and measure the service dimensions that are of primary importance to consumers, use the resulting factors to measure satisfaction levels with the service provided by operators in the electric vehicle market, and analyse consumer satisfaction when experiencing a problem such as one with service quality. In the present study, on the measurement of customer satisfaction with electric vehicle service and service quality, we identify and measure four dimensions - both the more obvious satisfaction dimensions already addressed in the marketing literature and new or confused issues by consumers. We also discuss the link between service satisfaction and complaint behaviour related to customer dissatisfaction after experiencing a service problem. The concept of service quality addresses a fit between consumer desires concerning service consumption and perceived service performance and receives attention once more. The emerging phenomena in the service industry, competition-driven concerns, and environmental consciousness also attract academic and commercial interest to examine the service quality issue in the electric vehicle market and evaluate its importance to customer satisfaction (Ali et al., 2021).

The continuous service quality improvement is essential in sustaining customer trust and developing loyalty. While product quality must be at least at baseline, organizations need to understand that customer satisfaction goes beyond the intrinsic quality of what is being offered. Therefore, the research objective is to investigate factors influencing drivers' satisfaction with EV service and quality. The research examines four key factors: service quality, vehicle quality, pricing, and customer experience.

The paper is organized as follows: The section 2 reviews the existing literature review related to topic after this introduction section. Section 3 describes the methodology used in the study while Section 4 presents the main findings of the study. The last section concludes the research with implications and limitations of the study.

2. LITERATURE REVIEW

Customer satisfaction is a key performance index for EV services, including for both the customers directly serviced by the EV charging stations and the customers who derive convenience from the stations because they are located nearby. A set of rules should be developed, derived from the principle that customer satisfaction is fundamental and that customer expectations are important. Quality is better defined by its dimensions in the service industry, and EV services seek to have many similarities with conventional services to maximize sustainability. Consequently, this study will modify different dimensions in the SERVQUAL questionnaire to adapt it to the EV setting. Numerous descriptions of the gap model already exist, with one application proposed by researchers. In their study, which considered both the enhancement of articulated measures and the exploration of new measures, six new dimensions were developed for the gap model: customization, e-tailment, contact, loyalty, trust, and value. Additional characteristics of service were also proposed, which we adapted for the EV service framework. The SERVQUAL model was applied to the EV setting to develop relevant dimensions and corresponding items. The work also lays the groundwork for the study for which the SERVQUAL model is adapted (Kwon et al., 2020).

Model predictors of customer expectations, consisting of the EV service setting and customer personal characteristics, were applied in this model. In view of the objective of the EV services, a service quality model for EV services was developed. The model consists of both the original five SERVQUAL dimensions and three new dimensions that were validated in EV settings. The original SERVQUAL dimensions tested are responsiveness, reliability, tangibles, assurance, and empathy. The three new dimensions of EV service quality are consistency, promptness, and location. EV service satisfaction is used as the criteria to evaluate the model. The proposed model is validated by two methods: exploratory factor analysis for reliability and confirmatory factor analysis for validity. The results show that EV services use many service cues and dimensions, including the SERVQUAL model dimensions and the three newly developed dimensions. Both perceived service quality and customer personal characteristics explain the level of experiential, normative, and affective purchase behaviours, such as post-purchase behaviours, customer loyalty, recommendations to others, and retention. Lastly, the effectiveness of promotion, advertising, and word-of-mouth suggestions will advance the development of green products. The future success of service innovation will rely on getting reliable and valid marketing transactions to customers by offering quality services at appropriate levels from multiple opportunities. Small markets and large markets for EVs and services should be developed because there are significantly different market segments for potential customers. Communicating the psychological benefit of purchasing an EV is also needed, such as investors and road-environment advocates, so that they can become less vulnerable to the influence of economic and financial risks (Ebrahimi et al., 2020).

2.1 Customer Satisfaction in the Automotive Industry

Among the traditional criteria governing the choice of automobiles and the assessment of car performance, customer satisfaction is noted to play a particularly important role in the automotive industry. There are a variety of reasons for this. On the demand side, customers are increasingly discerning in their tastes and personal expectations, and the dynamism of industrialized economies has led customers to pay greater attention to non-functional attributes of products such as satisfaction. Satisfaction is perceived, on the supply side, as playing an important role in determining loyalty and repeat purchases for consumers when buying a durable good, like a car, as opposed to a non-durable good. When customer satisfaction is lower, it needs to be made up by higher levels of customer satisfaction in maintenance and repair activities. Again, cars themselves are generally sold at a relatively low markup, compared to other products like clothes, and indeed in some cases manufacturers sell cars at tiny profit or even at a loss, offsetting any negative margin from sales of associated services (Balinado et al., 2021).

The view that customer satisfaction is of paramount importance in the car industry is supported by extensive economic literature based on concepts such as satisfaction, customer payback, and loyalty. Numerous research studies have indeed analyzed aspects that can influence car satisfaction, both at the level of pre-buying search and post-buying behavior. Preference studies and conjoint analysis have identified the attributes perceived as most important by customers in their choice of car, and the perception and actual experience arising during the possession and usage of a car have also been examined in detail. Factors significantly affecting satisfaction have been grouped into two main categories: influences (product) and contacts (services). The former play a fundamental role in shaping the customer's view of the product, who is interested in tangible elements such as the exterior and interior appearance of a car, engine power, safety, comfort, and equipment. The latter impact the process that fulfils a customer's need (search, purchase, use, etc.), and the satisfaction derived therefrom. Contacts are defined by the manner in which these elements are supplied by service and quality expectations. These are conditioned by the points of service that convey a product or service to the customer, and when the service process results in periodic contacts with a customer over time, service is conducive to customer satisfaction. Understanding the nature of these contacts facilitates the processing of high levels of customer satisfaction (Fu et al., 2020).

2.2 Service Quality and Customer Satisfaction

Service quality is viewed as a main determinant of customer satisfaction in the IS and marketing literature. Empirical studies in commercial services and IS provide strong support for the relationship between service quality and customer satisfaction. It is obvious that IS may affect satisfaction in a number of ways: they can be critical quality factors that directly contribute to the services; they can moderate or mediate the relationship between other quality factors and customer satisfaction; and high IS can act as delighters, that is, lead to unexpected levels of customer satisfaction when delivered at a certain quality (Marcos & Coelho, 2022).

In this study, we deliberately focus on customer satisfaction as the dependent variable, although it is very likely that service quality dimensions, and even IT-specific constructs such as system quality and service quality, are influential. Many of us often consider these conceptual models, which also include performance characteristics, that is, the organization's own service quality perceptions that reflect its ability to improve customer satisfaction directly. We also believe that SES does not need to be conceptually subsumed within the service quality dimensions, as this model highlights that an explicit focus on service quality enables us to compare different transport modes in terms of environmental quality, thereby making it more applicable to the EVS (Kwon et al., 2020).

2.3 Vehicle Quality and Customer Satisfaction

Besides service quality, vehicle quality also influences customer satisfaction with electric vehicles. Vehicle quality makes a difference, and whether the vehicle satisfies the owner may change over time. The quality of vehicles can shift relative to customer satisfaction, and customer satisfaction with electric vehicles gradually decreases over time. The level of customer satisfaction with newly acquired cars is higher than that of old cars, which shows that customer satisfaction with car service and lifecycle is indeed different. The sole technical indicators that can be inspected and proved about new energy vehicles might not fully reflect their early driver experience and customer satisfaction. Customer satisfaction with the automobile lifecycle will change, which is significant to adopt suitable policy tools for government and firms (Gulzari et al., 2022).

Even though it is not a vehicle quality inspection, a report measuring specific vehicle attributes and evaluating the driving experience of electric vehicle customers indicated that the non-technical attributes, such as visibility and roominess, contribute greatly to the highest scores. The customers reported that the electric vehicles were high quality relative to cost and economy and were easy to drive. The electric vehicle industry still faces problems, such as limited space in the car, limited driving range, limited charging infrastructure, restricted battery capacity, restricted appeal of electric vehicles, the battery effect, lack of support for external charging facilities, and adherence to purchasing patterns. The survey showed that the vehicle owner experience was affected by limited driving range, weak battery, and anxiety related to vehicle reliability, which might be an initial step to produce electric vehicle inconvenience and negative emotions. Car companies should release new model selection, lease, and rental plans for customers with lower trust and work to improve the charging infrastructure. In brief, government departments and firms should adopt suitable policy tools (Damaj et al., 2021).

2.4 Pricing Strategies and Customer Satisfaction

Economic factors can affect customer satisfaction with services through customer choice. Most scholars believe that consumers will choose companies that minimize costs and maximize services. Price is the main aspect of economic influence, and competitive pricing will provide high customer satisfaction. Some scholars believe that pricing can improve services such as in e-commerce, where low prices can attract new customers, and good quality brings customer satisfaction. Customer satisfaction has a significant impact on customer choice, such that customer satisfaction and customer choice will mutually affect each other. E-commerce provides an example of how low prices and good quality have strengthened the commitment between the customer and the brand, concerning both the customer side and the provider side in the e-commerce sportswear industry (Nguyen et al., 2020).

The sports, arts, and humanities industries are also deeply related to pricing. Companies continually face pressure to raise prices and offer discounts to sustain and attract demand. Research reveals that for most price-resistant products, prices must be raised marginally higher than discounts so that the right customers can be retained. It clearly defines the price of goods and also lists possible prices. They also think that marketers need to manage recognition to satisfy customer satisfaction. The study of the relation between earnings quality and customer satisfaction finds that companies that create more price transparency and effectively manage their negative context information, and thus assess the quality of their earnings, can bring more customer satisfaction. Reasons are also given about the prerequisites for understanding relationships. For instance, pricing and pricing methods will mainly affect customer satisfaction, and product design will also be secondary to packaging pricing. Different earning quality levels may show varying consumer perceptions of corporate costs, resulting in differences in customer satisfaction (Ali and Anwar, 2021).

2.5 Customer Experience and Satisfaction

Many authors argue that a positive experience is crucial to both satisfaction and loyalty, and the two concepts of customer satisfaction and service quality are closely related. However, the recognition to be made here is that although customer satisfaction is derived from expectations, and quality is used as the service evaluation, the range of customers' perception and participation in service have different effects on the factors that stimulate customer satisfaction. The literature has examined user satisfaction mainly from the perspective of service quality evaluation and implementation; the examination of the model under the perspective of different evaluation stimulus mechanisms is also an important supplementary area (Zouari & Abdelhedi, 2021).

The service evaluation perspective has already gained widespread acceptance in the service satisfaction literature. As a result, the core of customer satisfaction is attributed to the evaluation that follows a service encounter, while the interaction theory equally attributes customer satisfaction to the customer's perception and participation in that service encounter. In this perspective, the analysis of the relationship between corporate image and customer satisfaction on the part of academics essentially focuses on the comparison of the concepts, their dimensions, the attributes of goods themselves, and the related analysis of corporate brand names and impacts. This is carried out by examining the interplay of the concepts in question with customer loyalty. The above confirms that the level that customers perceive to be appropriate regarding service quality is an important stimulus to the determination of customer satisfaction (Ali et al., 2021).

3. DATA AND METHODOLOGY

This study made use of a structured questionnaire to gather information from 315 respondents and was a systematic way of understanding the relationship between demographic factors and key model variables: service quality, vehicle quality, pricing, customer experience, and customer satisfaction. The questionnaire gathered comprehensive demographic information, including age, gender, education, employment status, marital status, household income, region, and ownership status. These were essential in the contextualization of the findings and diversification within the sample population as well.

The analysis was conducted using SPSS, a reliable statistical software package, to perform descriptive analysis, reliability and validity checks, regression analysis, and post-regression diagnostic tests. The descriptive statistics summarized the demographic distribution, showing the representativeness and diversity of the sample. The reliability checks, such as Cronbach's alpha, ensured the consistency of the measurement scales, while validity checks confirmed the appropriateness of the data for further analysis.

The primary technique used was regression analysis, which tests for all relationships between independent variables - that are service quality, vehicle quality, pricing and customer experience, and the dependent variable - that is customer satisfaction. Significant predictors and their impact, thereby providing good information toward customer behaviour, were identified by using it.

The structured questionnaire ensured that data was collected in a consistent manner, and SPSS ensured comprehensive and accurate analysis. However, the study had limitations in terms of demographic imbalances, such as overrepresentation of urban respondents (86.7%) and male participants (64.1%). Such imbalances may limit the generalizability of the findings. In addition, the reliance on self-reported data introduces the possibility of response biases. Despite the above limitations, the proposed methodology served as a sound framework for analysing customer satisfaction, yet further research shall fill identified gaps and explore complementary qualitative methods.

4. RESULTS AND DISCUSSION

4.1 Descriptive Statistics of respondents

The Table 1 shows that most of the respondents are within the working ages of 25–34 years (37.8%) and 35–44 years (34.3%). The age groups reflect the middle age, which might indicate the population's active involvement in economic and social activities. The younger group of 18–24 years and the older group of 45+ years were not well represented, and it may be a limitation to the study as these populations are not well represented. The sample skews towards men, at 64.1%, while women are 30.8%. Industry-specific or survey-targeted demographics may skew the balance of this sample. However, it does draw attention to gender representation in the study and whether or not findings can be generalized to an equally represented population.

The educational profile indicates that a little more than half (52.7%) of these have a degree, whereas 20.6% had a master's degree, meaning the respondents are a very literate sample who will definitely guide their perceptions and responses. While the lower percentiles were only high school educated 15.2% or at the level of obtaining a doctorate 6.3%, there would be skewness in assessing the availability and access to tertiary education and subsequent socio-economic impact. The majority, 75.9%, are employed, indicating economic stability among respondents, while the unemployed group, 19%, provides a smaller but critical counterpoint to assess labour market challenges. The data implies a workforce-centric focus, potentially limiting the understanding of non-working demographics like students or retirees.

A large percentage (41.6%) report earning less than $20,000 a year, while smaller percentages indicate higher income brackets. Such a distribution reveals economic imbalances in the sample, enabling an understanding of financial limits that may influence consumer expenditure and other economic activities. The married population has 55.2% while singles stand at 39.7%. The difference suggests that answers from the married will have more to do with the family and household as opposed to those from singles.

Most respondents are urban dwellers with an urban-to-rural ratio at 86.7% versus 8.3%. As such, there could be differences in how service access, resources, and infrastructures might be deployed that would impact the research outcomes. There is a likelihood of results being inapplicable or limited in scope to rural residents given the urban dominance among respondents. Ownership data reveals that a higher percentage of ownership is present, at 54.3%, compared to leasing, at 40.6%.

Such a difference may indicate different levels of economic stability and long-term investment behaviour among respondents, which might be relevant to related studies on socioeconomic aspects.

Table 1. Demographics of respondents (Age)

Age					
		Frequency	**Percent**	**Valid Percent**	**Cumulative Percent**
Valid		16	5.1	5.1	5.1
	18-24	20	6.3	6.3	11.4
	25-34	119	37.8	37.8	49.2
	35-44	108	34.3	34.3	83.5
	45-54	37	11.7	11.7	95.2
	55-64	15	4.8	4.8	100.0
	Total	315	100.0	100.0	

Table 2. Demographics of respondents (Gender)

Gender					
		Frequency	**Percent**	**Valid Percent**	**Cumulative Percent**
Valid		16	5.1	5.1	5.1
	Female	97	30.8	30.8	35.9
	Male	202	64.1	64.1	100.0
	Total	315	100.0	100.0	

Table 3. Demographics of respondents (Education)

Education			
Frequency	**Percent**	**Valid Percent**	**Cumulative Percent**
16	5.1	5.1	5.1
166	52.7	52.7	57.8
20	6.3	6.3	64.1
48	15.2	15.2	79.4
65	20.6	20.6	100.0
315	100.0	100.0	

Table 4. Demographics of respondents (Employment Status)

Employment Status					
		Frequency	Percent	Valid Percent	Cumulative Percent
Valid		16	5.1	5.1	5.1
	Employed	239	75.9	75.9	81.0
	Unemployed	60	19.0	19.0	100.0
	Total	315	100.0	100.0	

Table 5. Demographics of respondents (Household income per year)

Household Income per year					
		Frequency	Percent	Valid Percent	Cumulative Percent
Valid		16	5.1	5.1	5.1
	$100,000 and above	20	6.3	6.3	11.4
	$20,000 - $39,999	43	13.7	13.7	25.1
	$40,000 - $59,999	60	19.0	19.0	44.1
	$60,000 - $79,999	29	9.2	9.2	53.3
	$80,000 - $99,999	16	5.1	5.1	58.4
	Less than $20,000	131	41.6	41.6	100.0
	Total	315	100.0	100.0	

Table 6. Demographics of respondents (Marital Status)

Marital Status					
		Frequency	Percent	Valid Percent	Cumulative Percent
Valid		16	5.1	5.1	5.1
	Married	174	55.2	55.2	60.3
	Single	125	39.7	39.7	100.0
	Total	315	100.0	100.0	

Table 7. Demographics of respondents (Region/Location)

Region/Location					
		Frequency	**Percent**	**Valid Percent**	**Cumulative Percent**
Valid		16	5.1	5.1	5.1
	Rural	26	8.3	8.3	13.3
	Urban	273	86.7	86.7	100.0
	Total	315	100.0	100.0	

Table 8. Demographics of respondents (Ownership Status)

Ownership Status					
		Frequency	Percent	Valid Percent	Cumulative Percent
Valid		16	5.1	5.1	5.1
	Lease	128	40.6	40.6	45.7
	Own	171	54.3	54.3	100.0
	Total	315	100.0	100.0	

4.2 Discussion on Reliability and Validity

The reliability analysis showed strong internal consistency with a Cronbach's Alpha value of 0.980 for the 24 items measured in the survey (Table 2). Such high value implies that the questionnaire items reliably measured the constructs they were meant to measure, thereby supporting the credibility of the data collected. Generally, a Cronbach's Alpha value above 0.7 is considered acceptable, whereas values above 0.9, as seen in this study, represent excellent reliability. This outcome further reiterates that the survey instrument was well-designed and effectively captured the nuances of the variables under study.

In terms of validity, the constructs were assessed through logical consistency and significant correlations among variables, ensuring that the items accurately represented the theoretical concepts they were designed to measure. The logical consistency of the constructs was evident in the alignment of the survey items with established frameworks and literature. Each item was mapped to its conceptual variable, such as the service quality, vehicle quality, pricing, customer experience, and customer satisfaction, respectively, ensuring conceptual clarity and relevance.

Table 9. Reliability test of items

Cronbach's Alpha	N of Items
.980	24

Significant correlations between variables further supported construct validity. For example, the relationship between customer experience and customer satisfaction was strong, in line with research that underlines the experiential factor as a significant determinant of overall satisfaction. In the same way, the relationship between service quality and pricing with customer satisfaction were expected, further supporting the validity of the constructs within the study's framework.

Although the reliability and validity metrics were sound, there should be some acknowledgement of potential issues that may arise. A high Cronbach's Alpha value, even though it reflects consistency, implies that there might be redundancy among items so that several items might be reflecting overlapping aspects of the same construct. Future versions of the instrument could consider how to reduce such redundancy without sacrificing breadth in construct coverage.

In conclusion, the assessments of validity and reliability form a very strong basis for the study findings, culminating in affirming the credibleness of the data and the appropriateness of the analytical framework. The results therefore guarantee the soundness of the insights derived from the analysis, mirroring these theoretical constructions.

4.3 Regression Analysis

Regression model explained 87.2% of the variance with an R-squared value of 0.872 (Table 3). Meaning that the predictors: service quality, vehicle quality, pricing, and customer experience explain 87.2% of the variation in customer satisfaction. This high value of R-squared indicates a robust explanatory capacity of the regression model.

Table 10. Model Summary

Model	R	R Square	Adjusted R Square	Std. Error of the Estimate	Durbin-Watson
1	.934[a]	.872	.871	1.24267	2.006
a. Predictors: (Constant), customerexperience, servicequality, pricing, vehiclequality					
b. Dependent Variable: customersatisfaction					

Furthermore, the Durbin-Watson statistic of 2.006 showed that there was no significant autocorrelation among the residuals, and thus, the independence assumption of the regression model was satisfied.

Table 11. ANOVA Analysis

Model		Sum of Squares	df	Mean Square	F	Sig.
1	Regression	3102.679	4	775.670	502.302	.000[b]
	Residual	454.003	294	1.544		
	Total	3556.682	298			
a. Dependent Variable: customersatisfaction						
b. Predictors: (Constant), customerexperience, servicequality, pricing, vehiclequality						

The ANOVA test further validated the overall significance of the model with a high F-statistic of 502.302 and a p-value < 0.001 (Table 4). This shows that, taken as a group, the predictors are significantly associated with customer satisfaction. The high F-statistic further indicates that the model is explanatory and robust, and therefore, the findings are valid.

Table 12. Results of multiple regression

		Unstandardized Coefficients		Standardized Coefficients		
Model		B	Std. Error	Beta	t	Sig.
1	(Constant)	.142	.372		.381	.703
	servicequality	.225	.060	.213	3.776	.000
	vehiclequality	.107	.048	.125	2.238	.026
	pricing	.384	.029	.589	13.126	.000
	customerexperience	.044	.038	.054	1.154	.249
a. Dependent Variable: customersatisfaction						

Coefficient and Predictor Analysis

Based on the regression analysis as shown in Table 5, service quality positively and significantly contributed to customer satisfaction (β=0.225, p <0.001). This is consistent with literature (Saidin et al., 2018; de Oña, 2022;) that points to service delivery as the most critical area that affects the perception and experience of customers. According to Saidin et al. (2018) service quality contributes to the positive relationship with customer loyalty. Quality service promotes trust and loyalty; thus, it remains an essential area for companies to pay attention to.

Vehicle quality also has a positive and significant influence on customer satisfaction ($\beta=0.107$, $p<0.05$). Our results are supported by findings of Waluya et al. (2019) and Balinado et al. (2021). Waluya et al. (2019) find that product quality influences customer satisfaction while Balinado et al, (2021) document that reliability of the vehicle has significant impact on satisfaction of customers. This means that though the product quality itself plays a role, it matters less than pricing and service quality. The customer probably expects an acceptable minimum quality level and going above or below it affects their level of satisfaction.

In the of pricing, it is also an influential predictor of customer satisfaction, with an unstandardized coefficient (β) of 0.384 and a p-value less than 0.001. Our findings are similar to the findings of Bhutto et al. (2022) and Hamzah & Pontes (2024). Bhutto et al. (2022) conclude that the price-sensitivity has been recognized as a key significant moderator in customer satisfaction. Meanwhile, Hamzah & Pontes (2024) reach that the perceived price fairness has significant impact on purchase intention of vehicles. This indicates that competitive and perceived fair pricing is strongly and directly associated with how satisfied customers feel. It highlights the significance of pricing strategy as a driver of customer satisfaction in markets with high-cost sensitivity.

Surprisingly, customer experience failed to achieve statistical significance regarding customer satisfaction ($\beta=0.044$, $p>0.05$). Our results are contradicted with findings of Yilmaz and Sürmelioğlu (2024) where they find customer experience has a significant and positive on customer satisfaction. This could imply that in this particular case, the experiential aspects are outweighed by other factors like price and service quality. Alternatively, it may also indicate that the measurement of customer experience is limited in the study or that the customer experience acts as a mediator instead of a direct variable.

5. CONCLUDING REMARKS

As main finding of the study, 1) service quality positively and significantly contributed to customer satisfaction ($\beta=0.225$, $p<0.001$); 2) Vehicle quality also has a positive and significant influence on customer satisfaction ($\beta=0.107$, $p<0.05$); 3) In the of pricing, it is also an influential predictor of customer satisfaction, with an unstandardized coefficient (β) of 0.384 and a p-value less than 0.001; last but not the least, customer experience failed to achieve statistical significance regarding customer satisfaction (β=0.044, p>0.05). The analysis pointed out pricing as the most significant determinant influencing customer satisfaction; it shapes perceptions and preferences. The second most important variable is service quality, which influences building trust and loyalty. Vehicle quality also had a positive impact, albeit smaller, indicating that product quality remains an essential component of customer satisfaction. Customer experience, while positively correlated with satisfaction, did not exhibit statistical significance, suggesting its potential indirect role or contextual limitations in this study.

The results provide actionable advice for EVs manufacturers. Competitive pricing strategies should be the first area of focus since they greatly improve customer satisfaction and retention. In the same way, continuous service quality improvement is essential in sustaining customer trust and developing loyalty. While product quality must be at least at baseline, organizations need to understand that customer satisfaction goes beyond the intrinsic quality of what is being offered. Sustained satisfaction is best achieved through investments in training, operational efficiency, and customer-centric practices.

The study has its own set of limitations. The sample mostly includes working, urban people. Such demography may not be representative of the population. Thus, a potential limitation of the findings will be the demographic bias or skewness of this kind, especially for a comparison to rural or diverse economies. Further, the cross-sectional study does not capture any dynamic elements or changes over time as far as customer preferences or satisfaction are concerned. The next research should focus on rectifying such limitations by including more varied demographics, possibly even longitudinal designs, for understanding the change in customer satisfaction. Also, the qualitative methods of interviewing or using a focus group might give a more intimate insight into customer expectations and preferences and supplement the findings in a quantitative manner. This can also be extended into how customer experience might function as a mediator or moderator within a satisfaction model. By addressing the identified limitations of the present study and extending its scope of research, future studies can build on the findings of these analyses to provide more nuanced details, which in turn would guide businesses in refining their strategies toward better customer engagement and retention.

REFERENCES

Ali, B. J., & Anwar, G. (2021). Marketing Strategy: Pricing strategies and its influence on consumer purchasing decision. *International journal of Rural Development. Environmental Health Report*, *5*(2), 26–39.

Ali, B. J., Gardi, B., Othman, B. J., Ahmed, S. A., Ismael, N. B., Hamza, P. A., & Anwar, G. (2021). Hotel service quality: The impact of service quality on customer satisfaction in hospitality. *International Journal of Engineering Business Management*, *5*(3), 14–28.

Aljarboa, S. (2024). Factors influencing the adoption of artificial intelligence in e-commerce by small and medium-sized enterprises. *International Journal of Information Management Data Insights*, *4*(2), 100285. DOI: 10.1016/j.jjimei.2024.100285

Balinado, J. R., Prasetyo, Y. T., Young, M. N., Persada, S. F., Miraja, B. A., & Redi, A. A. N. P. (2021). The effect of service quality on customer satisfaction in an automotive after-sales service. *Journal of Open Innovation*, *7*(2), 116. DOI: 10.3390/joitmc7020116

Bhutto, M. H., Tariq, B., Azhar, S., Ahmed, K., Khuwaja, F. M., & Han, H. (2022). Predicting consumer purchase intention toward hybrid vehicles: Testing the moderating role of price sensitivity. *European Business Review*, *34*(1), 62–84. DOI: 10.1108/EBR-10-2019-0274

Dam, S. M., & Dam, T. C. (2021). Relationships between service quality, brand image, customer satisfaction, and customer loyalty. *The Journal of Asian Finance, Economics and Business*, *8*(3), 585–593.

Damaj, I. W., Serhal, D. K., Hamandi, L. A., Zantout, R. N., & Mouftah, H. T. (2021). Connected and autonomous electric vehicles: Quality of experience survey and taxonomy. *Vehicular Communications*, *28*, 100312. DOI: 10.1016/j.vehcom.2020.100312

de Oña, J. (2022). Service quality, satisfaction and behavioral intentions towards public transport from the point of view of private vehicle users. *Transportation*, *49*(1), 237–269. DOI: 10.1007/s11116-021-10175-7

Ebrahimi, M., Rastegar, M., Mohammadi, M., Palomino, A., & Parvania, M. (2020). Stochastic charging optimization of V2G-capable PEVs: A comprehensive model for battery aging and customer service quality. *IEEE Transactions on Transportation Electrification*, *6*(3), 1026–1034. DOI: 10.1109/TTE.2020.3005875

Fu, X., Liu, S., Fang, B., Luo, X. R., & Cai, S. (2020). How do expectations shape consumer satisfaction? An empirical study on knowledge products. *Journal of Electronic Commerce Research, 21*(1), 1–20.

Goel, P., Sharma, N., Mathiyazhagan, K., & Vimal, K. E. K. (2021). Government is trying but consumers are not buying: A barrier analysis for electric vehicle sales in India. *Sustainable Production and Consumption, 28*, 71–90. DOI: 10.1016/j.spc.2021.03.029

Gulzari, A., Wang, Y., & Prybutok, V. (2022). A green experience with eco-friendly cars: A young consumer electric vehicle rental behavioral model. *Journal of Retailing and Consumer Services, 65*, 102877. DOI: 10.1016/j.jretconser.2021.102877

Hamzah, M. I., & Pontes, N. (2024). What drives car buyers to accept a rejuvenated brand? The mediating effects of value and pricing in a consumer-brand relationship. *Journal of Strategic Marketing, 32*(2), 114–136. DOI: 10.1080/0965254X.2022.2129749

Kwon, Y., Son, S., & Jang, K. (2020). User satisfaction with battery electric vehicles in South Korea. *Transportation Research Part D, Transport and Environment, 82*, 102306. DOI: 10.1016/j.trd.2020.102306

Marcos, A. M. B. D. F., & Coelho, A. F. D. M. (2022). Service quality, customer satisfaction and customer value: Holistic determinants of loyalty and word-of-mouth in services. *The TQM Journal, 34*(5), 957–978. DOI: 10.1108/TQM-10-2020-0236

Nguyen, D. T., Pham, V. T., Tran, D. M., & Pham, D. B. T. (2020). Impact of service quality, customer satisfaction and switching costs on customer loyalty. *The Journal of Asian Finance, Economics and Business, 7*(8), 395–405. DOI: 10.13106/jafeb.2020.vol7.no8.395

Saidin, Z. H., Mohd Mokhtar, S. S., Saad, R., & Yusoff, R. Z. (2018). The impact of automotive after-sales service quality and alternative attractiveness on customer loyalty. [IJSCM]. *International Journal of Supply Chain Management, 7*(1), 177–187.

Shokouhyar, S., Shokoohyar, S., & Safari, S. (2020). Research on the influence of after-sales service quality factors on customer satisfaction. *Journal of Retailing and Consumer Services, 56*, 102139. DOI: 10.1016/j.jretconser.2020.102139

Su, D., Gu, Y., Du, Q., Zhou, W., & Huang, Y. (2020). Factors affecting user satisfaction with new energy vehicles: A field survey in Shanghai and Nanjing. *Journal of Environmental Management, 270*, 110857. DOI: 10.1016/j.jenvman.2020.110857 PMID: 32721306

Waluya, A. I., Iqbal, M. A., & Indradewa, R. (2019). How product quality, brand image, and customer satisfaction affect the purchase decisions of Indonesian automotive customers. *International Journal of Services. Economics and Management, 10*(2), 177–193.

Yilmaz, V., & Sürmelİoğlu, Y. (2024), "Measuring automobile service quality with the European customer satisfaction index model (ECSI): the moderating effect of trust", *The TQM Journal*, Vol. ahead-of-print No. ahead-of-print. DOI: 10.1108/TQM-10-2023-0315

Zouari, G., & Abdelhedi, M. (2021). Customer satisfaction in the digital era: Evidence from Islamic banking. *Journal of Innovation and Entrepreneurship, 10*(1), 1–18. DOI: 10.1186/s13731-021-00151-x

Compilation of References

Abduljabbar, R., Dia, H., Liyanage, S., & Bagloee, S. A. (2019). The role of artificial intelligence in traffic signal control: A systematic review. *Journal of Transport and Land Use*, *12*(1), 163–194. DOI: 10.5198/jtlu.2019.1408

Abrahamse, W., & Steg, L. (2013). Social influence approaches to encourage resource conservation: A meta-analysis. *Global Environmental Change*, *23*(6), 1773–1785. DOI: 10.1016/j.gloenvcha.2013.07.029

Accenture. (2023). Monetizing digital services in automotive: A wake-up call for automakers to rethink their strategic priorities. https://www.accenture.com/us-en/insights/automotive/monetizing-digital-services-automotive

Acemoglu, D. (2002). Technical change, inequality, and the labor market. *Journal of Economic Literature*, *40*(1), 7–72. DOI: 10.1257/jel.40.1.7

Acemoglu, D., & Restrepo, P. (2019). Automation and new tasks: How technology displaces and reinstates labor. *The Journal of Economic Perspectives*, *33*(2), 3–30. DOI: 10.1257/jep.33.2.3

Aggarwal, S. (2024). Enhancing supply chain sustainability: The role of blockchain and digital technologies. *International Journal for Multidisciplinary Research*, *6*(5), 28451. Advance online publication. DOI: 10.36948/ijfmr.2024.v06i05.28451

Al-Hussaini, S., Al-Shehri, H., & Al-Zahrani, A. (2021). A review of AI-based solutions for smart parking management in urban areas. *Applied Sciences (Basel, Switzerland)*, *11*(13), 6100. DOI: 10.3390/app11136100

Ali, B. J., & Anwar, G. (2021). Marketing Strategy: Pricing strategies and its influence on consumer purchasing decision. *International journal of Rural Development. Environmental Health Report*, *5*(2), 26–39.

Ali, B. J., Gardi, B., Othman, B. J., Ahmed, S. A., Ismael, N. B., Hamza, P. A., & Anwar, G. (2021). Hotel service quality: The impact of service quality on customer satisfaction in hospitality. *International Journal of Engineering Business Management*, *5*(3), 14–28.

Aljarboa, S. (2024). Factors influencing the adoption of artificial intelligence in e-commerce by small and medium-sized enterprises. *International Journal of Information Management Data Insights*, *4*(2), 100285. DOI: 10.1016/j.jjimei.2024.100285

Alsaleh, T., & Farooq, B. (2024). Simulation models for sustainable, resilient, and optimized global electric vehicles supply chain. *arXiv*.

Alshurideh, M. T., Alzoubi, H. M., Kurdi, B. A., Hamadneh, S., Ahmed, G., Al-Sulaiti, K., Bataineh, A. Q., Alquqa, E. K., & Ozturk, I. (2025). Consumer and economic influences on electric vehicle adoption: The mediating role of attitudes and the moderating effect of demographics. *International Journal of Energy Economics and Policy*, *15*(3), 214–229. DOI: 10.32479/ijeep.19313

Álvarez, L., & González Marín, M. L. (Coords.). (2017). *Reestructuración productiva de la industria automotriz en la región del TLCAN, 2008–2015*. UNAM.

Amundsen, V. S. (2022). From checking boxes to actual improvement: A new take on sustainability certification. *Aquaculture (Amsterdam, Netherlands)*, *548*, 737672. DOI: 10.1016/j.aquaculture.2021.737672

Anastasiadou, K. (2021). Sustainable mobility driven prioritization of new vehicle technologies, based on a new decision-aiding methodology. *Sustainability (Basel)*, *13*(9), 4760. DOI: 10.3390/su13094760

Anderson, J. C., & Gerbing, D. W. (1988). Structural equation modeling in practice: A review and recommended two-step approach. *Psychological Bulletin*, *103*(3), 411–423. DOI: 10.1037/0033-2909.103.3.411

Andreoni, A., & Anzolin, G. (2025). GVCs meet robots: Smiling, smirking or flattening? The case of the automotive value chain. *. *Innovation and Development*, 1–23. Advance online publication. DOI: 10.1080/2157930X.2025.2468060

Andreoni, A., & Chang, H.-J. (2019). The political economy of industrial policy: Structural interdependencies, policy alignment and conflict management. *Structural Change and Economic Dynamics*, *48*, 136–150. DOI: 10.1016/j.strueco.2018.10.007

Anzolin, G. (2021). *Automation and labour in the age of Industry 4.0: The case of the automotive industry* (ICDD Working Paper).

Appolloni, A., Jabbour, C. J. C., D'Adamo, I., Gastaldi, M., & Settembre-Blundo, D. (2022). Green recovery in the mature manufacturing industry: The role of the green-circular premium and sustainability certification in innovative efforts. *Ecological Economics, 193*, 107311. DOI: 10.1016/j.ecolecon.2021.107311

Arteaga, A. (2020). El patrón de reproducción de capital exportador de especialización productiva y la industria automotriz en México. In Pinto, S., Guevara, A., & Arteaga, A. (Eds.), *La industria automotriz en América Latina: Estudios de las relaciones entre trabajo, tecnología y desarrollo socioeconómico** (pp. 129–160). Universidade Tecnológica Federal do Paraná.

Asociación Mexicana de la Industria Automotriz (AMIA). (2022). *El futuro de la electromovilidad en México*. AMIA.

Autor, D. H. (2015). Why are there still so many jobs? The history and future of workplace automation. *The Journal of Economic Perspectives, 29*(3), 3–30. DOI: 10.1257/jep.29.3.3

Autor, D. H., Levy, F., & Murnane, R. J. (2003). The skill content of recent technological change: An empirical exploration. *The Quarterly Journal of Economics, 118*(4), 1279–1333. DOI: 10.1162/003355303322552801

Balinado, J. R., Prasetyo, Y. T., Young, M. N., Persada, S. F., Miraja, B. A., & Redi, A. A. N. P. (2021). The effect of service quality on customer satisfaction in an automotive after-sales service. *Journal of Open Innovation, 7*(2), 116. DOI: 10.3390/joitmc7020116

Bamberg, S., & Möser, G. (2007). Twenty years after Hines, Hungerford, and Tomera: A new meta-analysis of psycho-social determinants of pro-environmental behaviour. *Journal of Environmental Psychology, 27*(1), 14–25. DOI: 10.1016/j.jenvp.2006.12.002

Barbarossa, C., & De Pelsmacker, P. (2014). Positive and negative antecedents of purchasing eco-friendly products: A comparison between green and non-green consumers. *Journal of Business Ethics, 134*(2), 229–247. DOI: 10.1007/s10551-014-2425-z

Barney, J. (1991). Firm resources and sustained competitive advantage. *Journal of Management, 17*(1), 99–120. DOI: 10.1177/014920639101700108

Barwick, P., Kwon, H., Li, S., Wang, Y., & Zahur, N. B. (2024, December 13). *Industrial policies and innovation in the electrification of the global automobile industry*. CEPR. https://cepr.org/voxeu/columns/industrial-policies-and-innovation-electrification-global-automobile-industry

Bem, D. J. (1972). Self-Perception Theory. *Advances in Experimental Social Psychology*, *6*, 1–62. DOI: 10.1016/S0065-2601(08)60024-6

Bensusán, G., & Gómez, W. (2017). Volkswagen de México: Un caso de articulación positiva y fuerte pero inestable entre innovación productiva y social. In J. Carrillo, G. Bensusán, & J. Micheli (Eds.), *¿Es posible innovar y mejorar laboralmente? Estudio de trayectorias de empresas multinacionales en México* (pp. 183–238). UAM Azcapotzalco.

Berman, E., Bound, J., & Griliches, Z. (1994). Changes in the demand for skilled labor within U.S. manufacturing: Evidence from the annual survey of manufactures. *The Quarterly Journal of Economics*, *109*(2), 367–397. DOI: 10.2307/2118467

Bhatnagar, K., & Sastry, C. (2020). Artificial intelligence and machine learning for sustainable transportation: A review. *Journal of Sustainable Development of Energy. Water and Environment Systems*, *8*(3), 478–499. DOI: 10.13044/j.sdewes.d7.0270

Bhutto, M. H., Tariq, B., Azhar, S., Ahmed, K., Khuwaja, F. M., & Han, H. (2022). Predicting consumer purchase intention toward hybrid vehicles: Testing the moderating role of price sensitivity. *European Business Review*, *34*(1), 62–84. DOI: 10.1108/EBR-10-2019-0274

Bleijenberg, A., Egenhofer, C., Behrens, A., Rizos, V., & Alessi, M. (2013). *Pathways to low carbon transport in the EU – From Possibility to reality*. Centre for European Policy Studies (CEPS). https://ssrn.com/abstract=2278809

Bloomberg, N. E. F. (2024). Electric Vehicle Outlook Report 2024. https://about.bnef.com/electric-vehicle-outlook/

Bohnsack, R., Pinkse, J., & Kolk, A. (2014). Business models for sustainable technologies: Exploring business model evolution in the case of electric vehicles. *Research Policy, 43*(2), 284–300. DOI: 10.1016/j.respol.2013.10.014

Bohnsack, R., Kolk, A., & Pinkse, J. (2020). Business models for sustainable technologies: Exploring business model evolution in the case of electric vehicles. *Research Policy*, *49*(2), 103–120. DOI: 10.1016/j.respol.2019.103928

Bonnefon, J., Shariff, A., & Rahwan, I. (2016). The social dilemma of autonomous vehicles. *Science*, *352*(6293), 1573–1576. DOI: 10.1126/science.aaf2654

Borsacchi, L. (2025). Certification Standards and Ecolabelling for Sustainability and Traceability in the Textile Supply Chain. In Sustainable Digital Marketing for Fashion and Luxury Brands: Theory and Practice (pp. 675-697). Cham: Springer Nature Switzerland. DOI: 10.1007/978-3-031-82467-8_28

Bown, C. P. (2023). *Industrial policy for electric vehicle supply chains and the US-EU fight over the Inflation Reduction Act | PIIE* (No. Industrial policy for electric vehicle supply chains and the US-EU fight over the Inflation Reduction). The Peterson Institute for International Economics. https://www.piie.com/publications/working-papers/2023/industrial-policy-electric-vehicle-supply-chains-and-us-eu-fight

Boyer, R., & Freyssenet, M. (2003). *Los modelos productivos*. Editorial Fundamentos.

Bueno-Pascual, F.-E. (2024). *Forces transforming transport and logistics into smarter sustainable systems*. IntechOpen. DOI: 10.5772/intechopen.1005001

Bukoff, A. (2020). Inside the Auto Industry: Widespread Skepticism about Electric Vehicles. SAE Technical Paper 2020-01-0529. DOI: 10.4271/2020-01-0529

Cao, H., Söderlund, H., Derspeisse, M., & Johansson, B. (2023). Exploring the current applications and potential of extended reality for environmental sustainability in manufacturing. *arXiv*.

Carlucci, F., Cirà, A., & Lanza, G. (2018). Hybrid electric vehicles: Some theoretical considerations on consumption behaviour. *Sustainability (Basel)*, *10*(4), 1302. DOI: 10.3390/su10041302

Carrillo, J., Martínez, A., López, O., & Díaz, M. (2023). The automotive sector in Mexico: The impact of automation and digitalization on employment. *Research Paper No. 15*. International Labour Organization.

Chang, H.-J., & Andreoni, A. (2016). *Industrial Policy in a Changing World: Basic Principles, Neglected Issues and New Challenges*. Cambridge Journal of Economics 40 Years Conference. https://www.semanticscholar.org/paper/Industrial-Policy-in-a-Changing-World%3A-Basic-Issues-Chang-Andreoni/e29246a6311db07a85006d33e4ac5cc6446cb90b?p2df

Chang, H.-J., & Andreoni, A. (2020). Industrial Policy in the 21st Century. *Development and Change*, *51*(2), 324–351. DOI: 10.1111/dech.12570

Chen, Z., Guo, S., & Li, Y. (2020). Sustainable traffic signal control using deep reinforcement learning with an attention mechanism. *IEEE Transactions on Intelligent Transportation Systems*, *21*(9), 3788–3798. DOI: 10.1109/TITS.2019.2934094

Chesbrough, H. W. (2003). *Open innovation: The new imperative for creating and profiting from technology*. Harvard Business School Press.

Chester, M. (2015). Can disruptive technologies, on-demand mobility, and biofuels improve transportation environmental sustainability? A review of recent research. DOI: 10.1007/s40518-015-0031-z

China Government Website. (2023, January 24). *China's New Energy Vehicle Production and Sales Rank First Globally for Eight Consecutive Years.* https://www.gov.cn/xinwen/2023-01/24/content_5738622.htm

Chirumalla, K., Klaff, S., Zako, R., & Sannö, A. (2023). Elevating B2B mobility with sharing autonomous electric vehicles: Exploring prerequisite criteria and innovative business models. *Sustainability (Basel), 15*(18), 13757. DOI: 10.3390/su151813757

Christensen, C., & Raynor, M. (2003). *The Innovator's Solution: Creating and Sustaining Successful Growth.* Harvard Business Review Press.

Chu, Y. (2023). *EV city profile: Passenger car market, Tianjin.* ICCT – International Council on Clean Transportation Europe. https://theicct.org/publication/ev-city-profile-passenger-car-market-tianjin/

Cialdini, R. B., & Goldstein, N. J. (2004). Social influence: Compliance and conformity. *Annual Review of Psychology, 55*(1), 591–621. DOI: 10.1146/annurev.psych.55.090902.142015 PMID: 14744228

Clinton, B. C., & Steinberg, D. C. (2019). Providing the Spark: Impact of financial incentives on battery electric vehicle adoption. *Journal of Environmental Economics and Management, 98*, 102255. DOI: 10.1016/j.jeem.2019.102255

Covarrubias, A. (2020). The boom of the Mexican automotive industry: From NAFTA to USMCA. In Covarrubias, A., & Ramírez, S. (Eds.), *New Frontiers of the Automobile Industry. Exploring Geographies, Technology, and Institutional Challenges** (pp. 323–348). Palgrave Macmillan. DOI: 10.1007/978-3-030-18881-8_13

Covarrubias, A., & Bouzas, A. (2016). *Empleo y políticas sindicales en la industria automotriz de México.* Friedrich Ebert Stiftung.

Daddi, T., Iraldo, F., & Testa, F. (2015). *Environmental Certification for Organisations and Products: Management approaches and operational tools.* Routledge. DOI: 10.4324/9781315768182

Damaj, I. W., Serhal, D. K., Hamandi, L. A., Zantout, R. N., & Mouftah, H. T. (2021). Connected and autonomous electric vehicles: Quality of experience survey and taxonomy. *Vehicular Communications, 28*, 100312. DOI: 10.1016/j.vehcom.2020.100312

Dam, S. M., & Dam, T. C. (2021). Relationships between service quality, brand image, customer satisfaction, and customer loyalty. *The Journal of Asian Finance, Economics and Business, 8*(3), 585–593.

De Buen, C. (2011). *Los contratos colectivos de trabajo de protección patronal en México. Cuadernos de Análisis Político.* Friedrich Ebert Stiftung.

de Oña, J. (2022). Service quality, satisfaction and behavioral intentions towards public transport from the point of view of private vehicle users. *Transportation, 49*(1), 237–269. DOI: 10.1007/s11116-021-10175-7

Degirmenci, K., & Breitner, M. H. (2017). Consumer purchase intentions for electric vehicles: Is green more important than price and range? *Transportation Research Part D, Transport and Environment, 51*, 250–260. DOI: 10.1016/j.trd.2017.01.001

Deloitte. (2022). The future of mobility: Transforming the automotive industry. https://www.deloitte.com

Deloitte. (2023). Sustainable supply chains and blockchain innovation in the automotive industry. https://www.deloitte.com

Deloitte. (2025). *2025 Global automotive consumer study: Key trends in mobility and sustainability*. Deloitte Insights. https://www2.deloitte.com/us/en/insights/industry/retail-distribution/global-automotive-consumer-study.html

Demirdöven, N., & Deutch, J. (2004). Hybrid cars now, fuel cell cars later. *Science, 305*(5686), 974–976. DOI: 10.1126/science.1093965 PMID: 15310893

Dewulf, J., Van Der Vorst, G., Denturck, K., Van Langenhove, H., Ghyoot, W., Tytgat, J., & Vandeputte, K. (2010). Recycling rechargeable lithium ion batteries: Critical analysis of natural resource savings. *Resources, Conservation and Recycling, 54*(4), 229–234. DOI: 10.1016/j.resconrec.2009.08.004

Dhanushya, J., Ruba, M., Ponkarthika, M., Soundarya, B., Sakthi Priya, V., Dhanushri, S., & Harikrishnan, S. (2025, April). AI-powered multipurpose agrobot for precision farming: Crop health management, fertilizer recommendation and yield prediction. In *Proceedings of the 2025 International Conference on Recent Advances in Electrical, Electronics, Ubiquitous Communication, and Computational Intelligence (RAEEUCCI)* (pp. 1–6). IEEE.

Dharmarathna, D., Graham, P., & Bunster, V. (2025). Unpacking the evolution of Green Star rating tools and their certification trends in Australian building projects. *Journal of Building Engineering, 113328*, 113328. Advance online publication. DOI: 10.1016/j.jobe.2025.113328

Dhivya, M., & Gowthami, S. (2020). AI-driven traffic management system for smart cities. *International Journal of Computer Engineering and Technology, 11*(4), 161–169. DOI: 10.34218/IJCET.11.4.2020.019

Dias, R., Irfan, M., Galvão, R. M., & Morão, A. (2024). Analyzing persistence in sustainable indices and global stock indices: Testing the market efficiency. In *Issues of sustainability in AI and new age thematic investing* (pp. 179–201). IGI Global. DOI: 10.4018/979-8-3693-3282-5.ch011

Díaz-García, C., González-Moreno, Á., & Sáez-Martínez, F. J. (2013). Gender diversity within R&D teams: Its impact on radicalness of innovation. *Innovation (North Sydney, N.S.W.)*, *15*(2), 149–160. DOI: 10.5172/impp.2013.15.2.149

Drucker, P. F. (1954). *The Practice of Management*. Heinemann.

Ebrahimi, M., Rastegar, M., Mohammadi, M., Palomino, A., & Parvania, M. (2020). Stochastic charging optimization of V2G-capable PEVs: A comprehensive model for battery aging and customer service quality. *IEEE Transactions on Transportation Electrification*, *6*(3), 1026–1034. DOI: 10.1109/TTE.2020.3005875

El Jaouhari, A., Arif, J., Samadhiya, A., Kumar, A., & Garza-Reyes, J. A. (2023). An environmental-based perspective framework: Integrating IoT technology into a sustainable automotive supply chain. *Benchmarking*. Advance online publication. DOI: 10.1108/BIJ-05-2023-0322

Elassy, M., Al-Hattab, M., Takruri, M., & Badawi, S. A. (2024). *Intelligent transportation systems for sustainable smart cities*. Transport Engineer. DOI: 10.1016/j.treng.2024.100252

Elkington, J. (1998). *Cannibals with forks: The triple bottom line of 21st century business*. Capstone.

Esiri, A. E., Kwakye, J. M., Ekechukwu, D., Ogundipe, O. B., & Ikevuje, A. H. (2023). Assessing the environmental footprint of the electric vehicle supply chain. *Magna Scientia Advanced Research and Reviews*, *8*(2), 219–227. DOI: 10.30574/msarr.2023.8.2.0099

Eurofound. (2018). *Automation, digitisation and platforms: Implications for work and employment*. Publications Office of the European Union.

European Commission. (2023). Aceleración del despliegue de las energías renovables. *European Commission*. https://ec.europa.eu/commission/presscorner/detail/es/ip_23_2061

Evans, P. (1995). *Embedded Autonomy: States and Industrial Transformation*. Princeton University Press., https://www.jstor.org/stable/j.ctt7t0sr DOI: 10.1515/9781400821723

Falchetta, G., & Noussan, M. (2021). Electric vehicle charging network in Europe: An accessibility and deployment trends analysis. *Transportation Research Part D, Transport and Environment, 94*, 102813. DOI: 10.1016/j.trd.2021.102813

Fishbein, M. & Ajzen, I. (2010). Predicting and changing behavior. DOI: 10.4324/9780203838020

Floridi, L., Cowls, J., Beltrametti, M., Chatila, R., Chazerand, P., Dignum, V., Luetge, C., Madelin, R., Pagallo, U., Rossi, F., Schafer, B., Valcke, P., & Vayena, E. (2018). AI4People—An Ethical Framework for a Good AI Society: Opportunities, risks, principles, and recommendations. *Minds and Machines, 28*(4), 689–707. DOI: 10.1007/s11023-018-9482-5

Fontaras, G., Pistikopoulos, P., & Samaras, Z. (2008). Experimental evaluation of hybrid vehicle fuel economy and pollutant emissions over real-world simulation driving cycles. *Atmospheric Environment, 42*(18), 4023–4035. DOI: 10.1016/j.atmosenv.2008.01.053

Fotis, G., Gupta, S., Varshney, T., Dhar, S. L., & Gulzar, M. M. (Eds.). (2025). *Modern Computing Technologies for EV Efficiency and Sustainable Energy Integration*. IGI Global. DOI: 10.4018/979-8-3373-2382-4

Fraga-Lamas, P., & Fernández-Caramés, T. M. (2024). A review on blockchain technologies for an advanced and cyber-resilient automotive industry. *arXiv*.

Frank, A. G., Dalenogare, L. S., & Ayala, N. F. (2019). Industry 4.0 technologies: Implementation patterns in manufacturing companies. *International Journal of Production Economics, 210*, 15–26. DOI: 10.1016/j.ijpe.2019.01.004

Freeman, R. B., & Medoff, J. L. (1984). *What do unions do?* Basic Books.

Fu, X., Liu, S., Fang, B., Luo, X. R., & Cai, S. (2020). How do expectations shape consumer satisfaction? An empirical study on knowledge products. *Journal of Electronic Commerce Research, 21*(1), 1–20.

Gao, S., Ma, X., & Li, H. (2020). A deep learning-based model for predicting traffic congestion in urban areas. *Journal of Big Data, 7*(1), 1–18. DOI: 10.1186/s40537-020-00336-7

García, H., Carrillo, J., & Bensusán, G. (Eds.). (2021). *Salarios en tiempos de libre comercio: ¿Ofrece la industria automotriz salarios dignos en México?* El Colegio de la Frontera Norte.

Garg, N., & Sharma, V. (2022). A review of AI and IoT-based solutions for smart and sustainable cities. *International Journal of Sustainable Engineering*, *15*(1), 1–13. DOI: 10.1080/19397038.2022.2030113

Geels, F. W. (2012). A socio-technical analysis of low-carbon transitions: Introducing the multi-level perspective into transport studies. *Journal of Transport Geography*, *24*(C), 471–482. DOI: 10.1016/j.jtrangeo.2012.01.021

Geels, F. W. (2019). Socio-technical transitions to sustainability: A review of criticisms and elaborations of the Multi-Level Perspective. *Current Opinion in Environmental Sustainability*, *39*, 187–201. DOI: 10.1016/j.cosust.2019.06.009

Ghasemi, H., & Zarrabi, M. (2022). Optimizing urban freight delivery using a multi-objective meta-heuristic algorithm: A case study with electric vehicles. *Journal of Cleaner Production*, *357*, 131976. DOI: 10.1016/j.jclepro.2022.131976

Gleim, M. R., Smith, J. S., Andrews, D., & Cronin, J. J.Jr. (2013). Against the Green: A multi-method examination of the barriers to green consumption. *Journal of Retailing*, *89*(1), 44–61. DOI: 10.1016/j.jretai.2012.10.001

Goel, P., Sharma, N., Mathiyazhagan, K., & Vimal, K. E. K. (2021). Government is trying but consumers are not buying: A barrier analysis for electric vehicle sales in India. *Sustainable Production and Consumption*, *28*, 71–90. DOI: 10.1016/j.spc.2021.03.029

Goldin, C., & Katz, L. F. (2008). *The race between education and technology*. Harvard University Press., DOI: 10.2307/j.ctvjf9vz

Gomes, A. D. P., Pauls, R., & ten Brink, T. (2023). Industrial policy and the creation of the electric vehicles market in China: Demand structure, sectoral complementarities and policy coordination. *Cambridge Journal of Economics*, *47*(1), 45–66. DOI: 10.1093/cje/beac056

Goos, M., & Manning, A. (2007). Lousy and lovely jobs: The rising polarization of work in Britain. *The Review of Economics and Statistics*, *8*(1), 118–133. DOI: 10.1162/rest.89.1.118

Goudarzi, G., & Mohammadi, A. (2021). AI applications in Hyperloop systems: A review on safety, efficiency, and sustainability. *Journal of Transportation Engineering. Part A, Systems*, *147*(9), 04021045. DOI: 10.1061/JTEAPL.0000620

Grand View Research. (2024). *Smart mobility market size and share report, 2024–2033*. https://www.grandviewresearch.com/industry-analysis/smart-mobility-market-report

Griffiths, K., Boyle, C., & Henning, T. F. (2018). Beyond the certification badge—How infrastructure sustainability rating tools impact on individual, organizational, and industry practice. *Sustainability (Basel), 10*(4), 1038. DOI: 10.3390/su10041038

Groening, C., Sarkis, J., & Zhu, Q. (2018). Green marketing consumer-level theory review: A compendium of applied theories and further research directions. *Journal of Cleaner Production, 172*, 1848–1866. DOI: 10.1016/j.jclepro.2017.12.002

Group, B. M. W. (2023). BMW Group report 2023. https://www.bmwgroup.com/en/report/2023/index.html

Group, C. A. R. (2021). The value chain is being replaced by the fast-changing automotive C.A.S.E. ecosystem (Connected, Autonomous, Shared, Mobility, Electrified). https://www.cargroup.org/wp-content/uploads/2021/08/Wakefield.pdf

Guanetti, J., Kim, Y., & Borrelli, F. (2018). Control of connected and automated vehicles: State of the art and future challenges. arXiv preprint. https://arxiv.org/abs/1804.03757

Gulzari, A., Wang, Y., & Prybutok, V. (2022). A green experience with eco-friendly cars: A young consumer electric vehicle rental behavioral model. *Journal of Retailing and Consumer Services, 65*, 102877. DOI: 10.1016/j.jretconser.2021.102877

Guo, Y., Shi, F., & Li, C. (2021). A comprehensive review of AI applications for sustainable transportation systems. *Sustainability, 13*(24), 13867. DOI: 10.3390/su132413867

Gupta, S. (2025). *The Future of Machine Learning with Automation in Engineering*. Nova Science Publishers. DOI: 10.52305/YFIX2832

Gupta, S., & Tripathi, R. K. (2014). Optimal LQR controller in CSC based STATCOM using GA and PSO. *Archives of Electrical Engineering, 63*(3), 469–487. Advance online publication. DOI: 10.2478/aee-2014-0034

Gupta, S., & Tripathi, R. K. (2015). Transient Stability Enhancement of Multimachine Power System Using Robust and Novel Controller Based CSC-STATCOM. *Advances in Power Electronics, 2015*(1), 626731. DOI: 10.1155/2015/626731

Hamzah, M. I., & Pontes, N. (2024). What drives car buyers to accept a rejuvenated brand? The mediating effects of value and pricing in a consumer-brand relationship. *Journal of Strategic Marketing, 32*(2), 114–136. DOI: 10.1080/0965254X.2022.2129749

Hangzhou Commercial Bureau. (2022). *Implementation Rules for Subsidies for Purchase of New Energy Vehicles in Hangzhou, 2022*. https://www.hangzhou.gov.cn/art/2022/6/30/art_812262_59060267.html?eqid=932dc7c90000f0f50000000464891e8e

Hangzhou Government. (2025, February 21). *Autonomous charging vehicles hit Hangzhou city's streets*. https://www.ehangzhou.gov.cn/2025-02/21/c_292710.htm

Hangzhou Passenger Vehicle Management Office. (2022). *Hangzhou license plate lottery winning rate*. http://hz.bendibao.com/live/20201123/96674.shtm

Hangzhou Statistics Bureau. (2022). Hangzhou. *Statistical Yearbook. Annuaire Statistique*, *2022*, •••. http://tjj.hangzhou.gov.cn/art/2022/11/28/art_1229453592_4107024.html

Hangzhou Urban and Rural Construction Committee. (2017). *Notice on the Issuance of Interim Measures for the Operation and Management of Public Charging Piles for New Energy Electric Vehicles in Hangzhou* (No. Hangzhou Construction Science and Technology Development [2017] No. 70). http://www.caam.org.cn/chn/9/cate_104/con_5207545.html

Hawkins, T. R., Singh, B., Majeau-Bettez, G., & Strømman, A. H. (2012). Comparative Environmental Life cycle assessment of conventional and electric vehicles. *Journal of Industrial Ecology*, *17*(1), 53–64. DOI: 10.1111/j.1530-9290.2012.00532.x

He, M. (2022, September 28). *Business Geography of China's NEV Cities⑦|Hangzhou: The First Internet City's "Buddhist" Competition*. https://www.21jingji.com/article/20220928/herald/f656da9b76eb8a547e7e5593b4d563ff.html

Heilmann, S., & Perry, E. J. (2020). *Mao's Invisible Hand*. Brill. https://brill.com/edcollbook/title/58480

Heilmann, S. (2018). *Red Swan: How Unorthodox Policy-Making Facilitated China's Rise*. The Chinese University of Hong Kong Press.

Heras-Saizarbitoria, I., Arana, G., & Boiral, O. (2015). Exploring the dissemination of environmental certifications in high and low polluting industries. *Journal of Cleaner Production*, *89*, 50–58. DOI: 10.1016/j.jclepro.2014.10.088

Higueras-Castillo, E., Guillén, A., Herrera, L. J., & Liébana-Cabanillas, F. (2020). Adoption of electric vehicles: Which factors are really important? *International Journal of Sustainable Transportation*, *15*(10), 799–813. DOI: 10.1080/15568318.2020.1818330

Higueras-Castillo, E., Kalinic, Z., Marinkovic, V., & Liébana-Cabanillas, F. J. (2019). A mixed analysis of perceptions of electric and hybrid vehicles. *Energy Policy*, *136*, 111076. DOI: 10.1016/j.enpol.2019.111076

Higueras-Castillo, E., Liébana-Cabanillas, F., Dos Santos, M. A., Zulauf, K., & Wagner, R. (2024). Do you believe it? Green advertising skepticism and perceived value in buying electric vehicles. *Sustainable Development (Bradford)*, *32*(5), 4671–4685. DOI: 10.1002/sd.2932

Higueras-Castillo, E., Rialp-Criado, J., Buhmann, K., & Liebana-Cabanillas, F. (2025). "Should I stay or should I go?" A Dual-Theory Behavioral Analysis of Electric vehicle adoption and Retention. *Transport Policy*, *103792*, 103792. Advance online publication. DOI: 10.1016/j.tranpol.2025.103792

Higueras-Castillo, E., Singh, V., Singh, V., & Liébana-Cabanillas, F. (2023). Factors affecting adoption intention of electric vehicle: A cross-cultural study. *Environment, Development and Sustainability*, *26*(11), 29293–29329. DOI: 10.1007/s10668-023-03865-y

Hirsch-Kreinsen, H., & Hompel, M. (2017). Digitalisierung industrieller Arbeit: Entwicklungsperspektiven und Gestaltungsansätz.

Hofstetter, G., Ma, Z., & Sgouridis, S. (2019). The role of Hyperloop in a sustainable transportation future: A techno-economic and environmental assessment. *Transportation Research Part D, Transport and Environment*, *69*, 312–327. DOI: 10.1016/j.trd.2019.02.008

Huang, K., Kockelman, K. M., & Gurumurthy, K. M. (2022). An overview of connected, automated, shared, and electric technologies. *Transportation Letters*, *14*(6), 529–540. DOI: 10.1080/19427867.2022.2059703

Huang, Y., Xu, J., & Zhang, H. (2019). Reinforcement learning for energy-efficient train operation in high-speed rail. *IEEE Transactions on Intelligent Transportation Systems*, *20*(9), 3469–3478. DOI: 10.1109/TITS.2018.2882252

Humphrey, J., & Memedovic, O. (2003). *The global automotive industry value chain: What prospects for upgrading by developing countries?* United Nations Industrial Development Organization (UNIDO). https://www.unido.org/resources/publications/cross-cutting-services/industrial-policy/global-automotive-industry-value-chain

IFR. (2024). Récord de 4 millones de robots en fábricas de todo el mundo. *Sala de prensa IFR*. https://ifr.org/ifr-press-releases/news/record-of-4-million-robots-working-in-factories-worldwide

International Energy Agency (IEA). (2023). CO_2 emissions from fuel combustion 2023 highlights. https://www.iea.org

International Energy Agency. (2024). *Global EV Outlook 2024.* https://www.iea.org/reports/global-ev-outlook-2024

IPCC. (2023). AR6 Synthesis Report. *IPCC.* https://www.ipcc.ch/report/sixth-assessment-report-cycle/

Irfan, M., Durrah, O., Chidambaram, N., & Singh, K. A. (2025). Green blockchain mechanisms: Enhancing transparency and efficiency in sustainable finance with DeFi solutions. In *Energy efficient deep learning approaches in IoT fog and green blockchain revolution* (pp. 211–232). IGI Global. DOI: 10.4018/979-8-3373-0300-0.ch008

Jamoussi, B., Abu-Rizaiza, A., & AL-Haij, A. (2022). Sustainable building standards, codes and certification systems: The status quo and future directions in Saudi Arabia. *Sustainability (Basel), 14*(16), 10314. DOI: 10.3390/su141610314

Johri, S., Selvakumar, P., Satyanarayana, P., Das, A., Gupta, S., & Bhattacharya, S. (2025). Urbanization and Energy: The Nexus of Global Challenges. In Geopolitical Landscapes of Renewable Energy and Urban Growth (pp. 453-478). IGI Global Scientific Publishing. DOI: 10.4018/979-8-3693-8814-3.ch016

Kagermann, H., Wahlster, W., & Helbig, J. (2013). *Recommendations for implementing the strategic initiative INDUSTRIE 4.0: Securing the future of German manufacturing industry* (Final report of the Industrie 4.0 Working Group). acatech – National Academy of Science and Engineering. https://www.acatech.de/wp-content/uploads/2018/03/Final_report__Industrie_4.0_accessible.pdf

Kahneman, D. (2011). *Thinking, Fast and Slow*. Penguin Books.

Kalafatis, S. P., Pollard, M., East, R., & Tsogas, M. H. (1999). Green marketing and Ajzen's theory of planned behaviour: A cross-market examination. *Journal of Consumer Marketing*, *16*(5), 441–460. DOI: 10.1108/07363769910289550

Kamaruddin, N., Osman, M. R., & Subari, K. (2013). Impact of robotics on employment and motivation of employees in the manufacturing sector. *International Journal of Engineering Research and Applications, 3*(2), 658–665.

Kameswaran, A., Nanthakumar, K., & Ruba, M. (2017). High performance precomputation based self-controlled precharge-free content-addressable memory. *Journal of Electronics and Communication Systems*, *2*, 1–9.

Kashem, M. A., Shamsuddoha, M., & Nasir, T. (2024). Sustainable transportation solutions for intelligent mobility: A focus on renewable energy and technological advancements for electric vehicles (EVs) and flying car. *Preprints*. DOI: 10.20944/preprints202405.1677.v1

Katz, L. F., & Autor, D. H. (1999). Changes in the wage structure and earnings inequality. In O. Ashenfelter & D. Card (Eds.), *Handbook of Labor Economics* (Vol. 3A, pp. 1463–1555). Elsevier. DOI: 10.1016/S1573-4463(99)03007-2

Katz, L. F., & Murphy, K. M. (1992). Changes in relative wages, 1963–1987: Supply and demand factors. *The Quarterly Journal of Economics*, *107*(1), 35–78. DOI: 10.2307/2118323

Kempton, W., & Tomić, J. (2005). Vehicle-to-grid power fundamentals: Calculating capacity and net revenue. *Journal of Power Sources*, *144*(1), 268–279. DOI: 10.1016/j.jpowsour.2004.12.025

Kern, F., Kivimaa, P., Rogge, K., & Rosenow, J. (2018). Policy mixes for sustainable energy transitions: The case of energy efficiency. In *Transitions in Energy Efficiency and Demand*. Routledge. DOI: 10.4324/9781351127264-12

Kern, F., Rogge, K. S., & Howlett, M. (2019). Policy mixes for sustainability transitions: New approaches and insights through bridging innovation and policy studies. *Research Policy*, *48*(10), 103832. DOI: 10.1016/j.respol.2019.103832

Keynes, J. M. (1930). Economic possibilities for our grandchildren. In *Essays in Persuasion* (pp. 358–373). W. W. Norton.

Kline, R. B. (2016). *Principles and practice of structural equation modeling* (4th ed.). The Guilford Press.

Klöckner, C. A., & Blöbaum, A. (2010). A comprehensive action determination model: Toward a broader understanding of ecological behaviour using the example of travel mode choice. *Journal of Environmental Psychology*, *30*(4), 574–586. DOI: 10.1016/j.jenvp.2010.03.001

Klöckner, C. A., & Matthies, E. (2009). Structural modeling of car use on the way to the university in different settings: Interplay of norms, habits, situational restraints, and perceived behavioral control. *Journal of Applied Social Psychology*, *39*(8), 1807–1834. DOI: 10.1111/j.1559-1816.2009.00505.x

Klöckner, C. A., Matthies, E., & Hunecke, M. (2003). Problems of operationalizing habits and integrating habits in normative Decision-Making models. *Journal of Applied Social Psychology*, *33*(2), 396–417. DOI: 10.1111/j.1559-1816.2003.tb01902.x

Klooster, D. (2006). Environmental certification of forests in Mexico: The political ecology of a nongovernmental market intervention. *Annals of the Association of American Geographers, 96*(3), 541–565. DOI: 10.1111/j.1467-8306.2006.00705.x

Kollmuss, A., & Agyeman, J. (2002). Mind the Gap: Why do people act environmentally and what are the barriers to pro-environmental behavior? *Environmental Education Research, 8*(3), 239–260. DOI: 10.1080/13504620220145401

Koroma, M. S., Costa, D., Philippot, M., Cardellini, G., Hosen, M. S., Coosemans, T., & Messagie, M. (2022). Life cycle assessment of battery electric vehicles: Implications of future electricity mix and different battery end-of-life management. *The Science of the Total Environment, 831*, 154859. DOI: 10.1016/j.scitotenv.2022.154859 PMID: 35358517

KPMG International. (2024). *Supply chain trends 2024: The digital shake-up*. https://kpmg.com/xx/en/our-insights/ai-and-technology/supply-chain-trends-2024.html

KPMG. (2023). Digital transformation in automotive: Driving value in the connected era. https://home.kpmg

KPMG. (2024, September). The future of automotive supply chains: Connected, collaborative and resilient. *KPMG Insights*. https://kpmg.com/in/en/blogs/2024/09/the-future-of-automotive-supply-chains-connected-collaborative-and-resilient.html

Krishnan, R., Phan, P. Y., Krishnan, S. N., Agarwal, R., & Sohal, A. (2024). Industry 4.0-driven business model innovation for supply chain sustainability: An exploratory case study. *Business Strategy and the Environment*. Advance online publication. DOI: 10.1002/bse.3970

Krithika, S., Jayachandran, T., Raghapriya, N. R., Ruba, M., Janani, C., & Sathiyavathi, S. (2025, April). An Effective Denial of Service Texture Features from Malware Grayscale Image. In 2025 3rd International Conference on Artificial Intelligence and Machine Learning Applications Theme: Healthcare and Internet of Things (AIMLA) (pp. 1-7). IEEE. DOI: 10.1109/AIMLA63829.2025.11041501

Krzywdzinski, M. (2017). Automation, skill requirements and labour-use strategies: High-wage and low-wage approaches to high-tech manufacturing in the automotive industry. *New Technology, Work and Employment, 32*(3), 247–267.

Kuiper, G., & Gemählich, A. (2017). Sustainability and depoliticisation: Certifications in the cut-flower industry at Lake Naivasha, Kenya. *Africa Spectrum, 52*(3), 31–53. DOI: 10.1177/000203971705200302

Kwon, Y., Son, S., & Jang, K. (2020). User satisfaction with battery electric vehicles in South Korea. *Transportation Research Part D, Transport and Environment*, *82*, 102306. DOI: 10.1016/j.trd.2020.102306

Lampón, J. F., Cabanelas-Lorenzo, P., & Delgado Guzmán, J. A. (2018). Claves en la evolución de México dentro de la cadena de valor global de la industria de autopartes: El caso del Bajío. *El Trimestre Economico*, *85*(339), 483–514. DOI: 10.20430/ete.v85i339.259

Leonidou, L. C., Coudounaris, D. N., Kvasova, O., & Christodoulides, P. (2015). Drivers and Outcomes of Green Tourist Attitudes and Behavior: Sociodemographic Moderating Effects. *Psychology and Marketing*, *32*(6), 635–650. DOI: 10.1002/mar.20806

Li, G., Yu, W., Chen, X., Wang, W., Guo, H., & Jiang, X. (2024). Exploring User Acceptance of Autonomous Vehicles: Impact of Driver and Vehicle Styles. *SAE International Journal of Connected and Automated Vehicles, 8*, 193-201. DOI: 10.4271/12-08-02-0015

Li, G., Song, M., & Sun, Y. (2020). AI-based predictive maintenance for urban rail transit systems. *IEEE Transactions on Industrial Informatics*, *16*(3), 1957–1965. DOI: 10.1109/TII.2019.2929427

Lima, L., Trindade, E., Alencar, L., Alencar, M., & Silva, L. (2021). Sustainability in the construction industry: A systematic review of the literature. *Journal of Cleaner Production*, *289*, 125730. DOI: 10.1016/j.jclepro.2020.125730

Liu, Y., Li, J., & Yang, B. (2021). A multi-agent reinforcement learning approach for green traffic signal control. *IEEE Transactions on Systems, Man, and Cybernetics. Systems*, *51*(8), 4991–5001. DOI: 10.1109/TSMC.2019.2946272

Liu, Z., & Xie, T. (2024). The Impact of Policy Thematic Differences on Industrial Development: An Empirical Study Based on China's Electric Vehicle Industry Policies at the Central and Local Levels. *Energies*, *17*(22), 22. Advance online publication. DOI: 10.3390/en17225805

Lopes, J. P., Soares, F. J., & Almeida, P. M. R. (2010). Integration of electric vehicles in the electric power system. *Proceedings of the IEEE*, *99*(1), 168–183. DOI: 10.1109/JPROC.2010.2066250

Luo, J., & Zhang, W. (2019). Carbon emission reduction in a logistics network using an AI-based vehicle routing algorithm. *Journal of Cleaner Production*, *222*, 974–984. DOI: 10.1016/j.jclepro.2019.03.040

Lu, S., & Wang, H. (2022). Limited Decentralization: Understand China's Land System from the Perspective of Central-Local Relation. *Land (Basel), 11*(4), 4. Advance online publication. DOI: 10.3390/land11040517

Magnusson, T., & Werner, V. (2023). Conceptualisations of incumbent firms in sustainability transitions: Insights from organisation theory and a systematic literature review. *Business Strategy and the Environment, 32*(2), 903–919. DOI: 10.1002/bse.3081

Malcalm, E., Asiedu, E., Irfan, M., & Awudu, I. (2024). Strengthening economic relationships and service supplier engagement: Pathways to indigenous inclusivity and sustainable development. In *Future of indigenous inclusivity: Unlocking finance for economic development* (pp. 129–149). IGI Global. DOI: 10.4018/979-8-3693-7924-0.ch007

Marchi, A. (2022). Environmental sustainability certifications in the Fashion Industry.

Marcos, A. M. B. D. F., & Coelho, A. F. D. M. (2022). Service quality, customer satisfaction and customer value: Holistic determinants of loyalty and word-of-mouth in services. *The TQM Journal, 34*(5), 957–978. DOI: 10.1108/TQM-10-2020-0236

MarkWide Research. (2023, February). Global connected, autonomous, shared, and electric (CASE) market set for remarkable growth, predicted to reach $1.6 trillion by 2030. Press release.

Marques, N., Campos, G., & Cavalcante, M. (2024). Driving towards a sustainable future: Leveraging connected vehicle data for effective carbon emission management. *SAE Technical Paper Series*. DOI: 10.4271/2023-36-0145

Masoumi, S. M., Kazemi, N., & Abdul-Rashid, S. H. (2019). Sustainable supply chain management in the automotive industry: A process-oriented review. *Sustainability (Basel), 11*(14), 3945. DOI: 10.3390/su11143945

Mazar, N., & Zhong, C. (2010). Do green products make us better people? *Psychological Science, 21*(4), 494–498. DOI: 10.1177/0956797610363538 PMID: 20424089

Mazzocco, I. (2025). *Electric Vehicles, China, and the Industrial Strategies Reshaping Mobility in Emerging Economies*. Center for Strategic and International Studies. https://www.csis.org/analysis/electric-vehicles-china-and-industrial-strategies-reshaping-mobility-emerging-economies

Mazzucato, M. (2016). From market fixing to market-creating: A new framework for innovation policy. *Industry and Innovation*. https://www.tandfonline.com/doi/abs/10.1080/13662716.2016.1146124

McKinsey & Company. (2022). The road to seamless mobility: Rethinking ecosystems in the automotive sector. https://www.mckinsey.com

McKinsey & Company. (2025, April 22). *New twists in the electric-vehicle transition: A consumer perspective*. https://www.mckinsey.com/features/mckinsey-center-for-future-mobility/our-insights/new-twists-in-the-electric-vehicle-transition-a-consumer-perspective

Méndez, M. (2025, January 8). La industria automotriz en México: Plantas armadoras de vehículos y su distribución estratégica. *Mexico Industry*. https://mexicoindustry.com/noticia/la-industria-automotriz-en-mexico-plantas-armadoras-de-vehiculos-y-su-distribucion-estrategica

Messagie, M. (2014). Life cycle analysis of the climate impact of electric vehicles. *Journal of Life Cycle Assessment*, 14. https://scholar.google.be/citations?view_op=view_citation&hl=nl&user=3qpRhxwAAAAJ&citation_for_view=3qpRhxwAAAAJ:hFOr9nPyWt4C

Middlebrook, K. (1988). Industrial restructuring and union politics in the Mexican automobile industry. Paper presented at the conference *México y la Cuenca del Pacífico*.

Ministry of Finance. (2021). *Notice on the Financial Subsidy Policy for the Popularization and Application of New Energy Vehicles in 2022* (No. Financial Construction [2021] No. 466). https://www.gov.cn/zhengce/zhengceku/2021-12/31/content_5665857.htm

Mishra, S., & Garg, P. (2022). A deep learning approach for traffic flow prediction in a smart city environment. *Journal of Ambient Intelligence and Humanized Computing*, *13*(5), 2379–2391. DOI: 10.1007/s12652-021-03004-8

Mitropoulos, L. K., Prevedouros, P. D., & Kopelias, P. (2017). Total cost of ownership and externalities of conventional, hybrid and electric vehicle. *Transportation Research Procedia*, *24*, 267–274. DOI: 10.1016/j.trpro.2017.05.117

MoF, & MCT. (2009). *Notice on the development of energy saving and NEVs Demonstration and Promotion Pilot Program* (No. Finance [2009] No. 6). https://www.gov.cn/zwgk/2009-02/05/content_1222338.htm

Mohanty, S., Kumar, S., & Dash, R. R. (2022). AI-driven smart infrastructure for sustainable transportation in smart cities. In Dash, R. R., & Mohanty, S. S. (Eds.), *Smart Cities and Urban Development* (pp. 235–256). Wiley., DOI: 10.1002/9781119792015.ch10

Moons, I., & De Pelsmacker, P. (2015). An extended decomposed theory of planned behaviour to predict the usage intention of the electric car: A Multi-Group Comparison. *Sustainability (Basel)*, *7*(5), 6212–6245. DOI: 10.3390/su7056212

Moore, G. A. (1991). *Crossing the Chasm: Marketing and Selling High-Tech Goods to Mainstream Customers*. Harper Business.

Moradloo, N., Mahdinia, I., & Khattak, A. (2024). Charging into the Future: Unraveling the Factors Shaping Electric Vehicle Adoption and Addressing Heterogeneity. *SAE International Journal of Sustainable Transportation, Energy, Environment, & Policy, 6*, 51-63. .DOI: 10.4271/13-06-01-0005

Morales, J. J., & López, M. G. (2018). Honda y su mercado de trabajo en la planta de El Salto, Jalisco. In de la Garza, E., & Hernández, R. (Eds.), *Configuraciones productivas y laborales en la tercera generación de la industria automotriz terminal en México** (pp. 131–157). UAM / Miguel Ángel Porrúa.

Moreno-Brid, J. C., Gómez Tovar, R., Sánchez Gómez, J., & Gómez Rodríguez, L. (2021). Trade agreements and decent work in Mexico: The case of the automobile and textile industries. *ILO Working Paper No. 36*. International Labour Organization.

Neckermann Strategic Advisors. (2020). The future mobility value chain: Autonomous mobility, the disrupting piece of the complex puzzle. Medium. https://medium.com/@Neckermann.Strategic/the-future-mobility-value-chain-autonomous-mobility-the-disrupting-piece-of-the-complex-puzzle-c42c0fa0ab4d

Netsol Tech. (2024). CASE (Connected, Autonomous, Shared and Electric): An insight. https://netsoltech.com/blog/case-connected-autonomous-shared-and-electric-an-insight

Nguyen, D. T., Pham, V. T., Tran, D. M., & Pham, D. B. T. (2020). Impact of service quality, customer satisfaction and switching costs on customer loyalty. *The Journal of Asian Finance, Economics and Business*, *7*(8), 395–405. DOI: 10.13106/jafeb.2020.vol7.no8.395

OECD. (2020). *The digitalisation of science, technology and innovation: Key developments and policies*. https://doi.org/DOI: 10.1787/b9e4a2c7-en

OECD. (2023). AI in transportation systems policy frameworks. IEA. (2024). AI for energy and mobility integration.

OICA. (2022). 2021 *Production statistics. International Organization of Motor Vehicle Manufacturers*. http://www.oica.net/category/production-statistics/2019-statistics/

Olson, E. (2012). It's not easy being green: The effects of attribute tradeoffs on green product preference and choice. *Journal of the Academy of Marketing Science, 41*(2), 171–184. DOI: 10.1007/s11747-012-0305-6

Omar, N. H., Mohd, R. H., & Syaima, N. N. (2025). The impact of automation adoption on employment structures: A longitudinal panel analysis of skill-biased technological change in Malaysia's manufacturing sector, 2012–2023. *International Journal of Academic Research in Progressive Education and Development, 14*(2), 417–431. DOI: 10.6007/IJARPED/v14-i2/25260

Orsato, R., & Wells, P. (2007). The Automobile Industry & Sustainability. *Journal of Cleaner Production, 15*(11–12), 989–993. DOI: 10.1016/j.jclepro.2006.05.035

Østergaard, C. R., Timmermans, B., & Kristinsson, K. (2011). Does a different view create something new? The effect of employee diversity on innovation. *Research Policy, 40*(3), 500–509. DOI: 10.1016/j.respol.2010.11.004

Ottman, J. A. (1998). *Green marketing: opportunity for innovation.* NTC Business Books.

Peattie, K. (1999). Trappings versus substance in the greening of marketing planning. *Journal of Strategic Marketing, 7*(2), 131–148. DOI: 10.1080/096525499346486

Peattie, K., & Crane, A. (2005). Green marketing: Legend, myth, farce or prophesy? *Qualitative Market Research, 8*(4), 357–370. DOI: 10.1108/13522750510619733

Pérez-Moure, H., Lampón, J. F., Velando-Rodriguez, M.-E., & Rodríguez-Comesaña, L. (2023). Revolutionizing the road: How sustainable, autonomous, and connected vehicles are changing digital mobility business models. *European Research on Management and Business Economics, 29*(3), 100230. DOI: 10.1016/j.iedeen.2023.100230

Poponi, S., Colantoni, A., Cividino, S. R., & Mosconi, E. M. (2019). The stakeholders' perspective within the B corp certification for a circular approach. *Sustainability (Basel), 11*(6), 1584. DOI: 10.3390/su11061584

Popovich, N. (2024). Where E.V.s are (and aren't) taking off across the U.S. *The New York Times.* https://www.nytimes.com/interactive/2024/03/06/climate/hybrid-electric-vehicle-popular.html

Porter, M. E., & van der Linde, C. (1995, September-October). Green and Competitive: Ending the Stalemate. *Harvard Business Review*, 119–134.

Prell, M., Zanini, M. T., Caldieraro, F., & Migueles, C. (2020). Sustainability certifications and product preference. *Marketing Intelligence & Planning*, *38*(7), 893–906. DOI: 10.1108/MIP-12-2019-0616

PRNewswire. (2023). S&P Global Mobility Survey Finds EV Affordability tops Charging and Range Concerns in Slowing EV Demand. *S&P Global Mobility*. https://press.spglobal.com/2023-11-08-S-P-Global-Mobility-Survey-Finds-EV-Affordability-tops-Charging-and-Range-Concerns-in-Slowing-EV-Demand

Prothero, A. (1996). Environmental decision making: Research issues in the cosmetics and toiletries industry. *Marketing Intelligence & Planning*, *14*(2), 19–25. DOI: 10.1108/02634509610110769

Puga Garcia, A. (2023, October 23). Towards a new connected, autonomous, shared and electric mobility. MAPFRE Insights. https://www.mapfre.com/en/insights/innovation/connected-autonomous-shared-electric-mobility/

PwC. (2023). Automotive trends 2023: Driving digital and sustainable transformation. https://www.pwc.com

Quiroga, R. Q. (2012). Borges and memory. In *The MIT Press eBooks*. DOI: 10.7551/mitpress/9396.001.0001

Richard, R.-B. (2005). Industrialised building systems: Reproduction before automation and robotics. *Automation in Construction*, *14*(4), 442–451. DOI: 10.1016/j.autcon.2004.09.009

Riofrío, C. A., Klepsch, J., Finžgar, J. R., Kiwit, F., Hölscher, L., Erdmann, M., Müller, L., Kumar, C., & Achari Berrada, Y. (2024). Quantum computing for automotive applications. *arXiv*.

Rodríguez, S., & Oropeza, A. (2024, September 2). Toyota revoluciona el futuro de la industria automotriz: Conoce el proceso de fabricación de su planta en Guanajuato. *Mexico Industry*. https://mexicoindustry.com/noticia/toyota-revoluciona-el-futuro-de-la-industria-automotriz-conoce-el-proceso-de-fabricacion-de-su-planta-en-guanajuato

Rodríguez-García, R., Ferrero-Ferrero, I., & Fernández-Izquierdo, M. Á. (2023). Analysis of integration of sustainability in sustainability certifications in the hotel industry. *Frontiers in Sustainability*, *4*, 1116359. DOI: 10.3389/frsus.2023.1116359

Rogers, E. M. (1962). *Diffusion of Innovations*. Free Press of Glencoe.

Rogge, K. S., & Reichardt, K. (2016). Policy mixes for sustainability transitions: An extended concept and framework for analysis. *Research Policy*, *45*(8), 1620–1635. DOI: 10.1016/j.respol.2016.04.004

Rosenstein-Rodan, P. N. (1961). Notes on the Theory of the 'Big Push.' In H. S. Ellis (Ed.), *Economic Development for Latin America: Proceedings of a Conference held by the International Economic Association* (pp. 57–81). Palgrave Macmillan UK. DOI: 10.1007/978-1-349-08449-4_3

Ruba, M., Ponkarthika, M., Parameshwari, V., Girinath, V. P., Soundarya, B., Janani, M., & Santhini, C. (2025, February). Enhancing underwater tunnel safety through smart inspection: An ROV-based system with AI-powered crack detection using Canny edge detection. In Proceedings of the 2025 3rd International Conference on Intelligent Systems, Advanced Computing and Communication (ISACC) (pp. 1067–1072). IEEE. DOI: 10.1109/ISACC.2025

Ruba, M., Ashwin Prabhu, G., Deka, J., Kavitha, D., Sathiya, S., & Pavithra, S. (2026). Empowering learners and educators: The transformative role of conversational AI in personalized education and student support. In Assayed, S. (Ed.), *Enhancing student support and learning through conversational AI* (pp. 163–190). IGI Global Scientific Publishing. DOI: 10.4018/979-8-3373-3316-8.ch007

Ruba, M., Ashwin Prabhu, G., Robinson Joel, M., Vishnu Priya, K., Dharshani, L., & Dharani, R. (2026). Integrating emerging technologies for organizational agility: An interdisciplinary approach to digital transformation. In Roache, D. (Ed.), *High performance leadership for organizational excellence: Innovation, technology, and resilience* (pp. 63–92). IGI Global Scientific Publishing. DOI: 10.4018/979-8-3373-6207-6.ch003

Ruba, M., Girinath, V. P., Ponkarthika, M., Santhini, C., Sakthi Priya, V., & Hemalatha, R. (2023, December). Design and implementation of smart internet-based real-time control of rover with artificial intelligence. In *Proceedings of the 2023 International Conference on Energy, Materials and Communication Engineering (ICEMCE)* (pp. 1–5). IEEE. DOI: 10.1109/ICEMCE2023

S&P Global. (2025, July). *2025 connected car study: Consumer priorities on connected services*. https://www.spglobal.com/automotive-insights/en/blogs/2025/07/connected-car-study-consumer-priorities-on-connected-services

Saadi, A., Abghour, N., Chiba, Z., Moussaid, K., & Ali, S. (2023). A survey of Reinforcement and Deep reinforcement learning for Coordination in Intelligent Traffic Light Control. *Reseach Square*. DOI: 10.21203/rs.3.rs-3225879/v1

Saidin, Z. H., Mohd Mokhtar, S. S., Saad, R., & Yusoff, R. Z. (2018). The impact of automotive after-sales service quality and alternative attractiveness on customer loyalty. [IJSCM]. *International Journal of Supply Chain Management, 7*(1), 177–187.

Santhini, C., Ruba, M., Sakthi Priya, V., & Hemalatha, R. (2025). Intercultural communication and assessment practices in the integration of AI with education. In *Generative AI Approaches to Sustainable Development in Higher Education*. IGI Global. DOI: 10.4018/979-8-3693-5623-4.ch010

Sarkar, D., & Pal, S. (2020). AI-enabled fleet management for carbon footprint reduction in last-mile delivery. *Procedia Computer Science, 171*, 1642–1651. DOI: 10.1016/j.procs.2020.06.168

Schultz, P. W. (2022). Secret Agents of Influence: Leveraging Social Norms for Good. *Current Directions in Psychological Science, 31*(5), 443–450. DOI: 10.1177/09637214221109572

Schumpeter, J. A. (1942). *Capitalism, socialism and democracy*. Harper & Brothers.

Schwab, K. (2016). *The Fourth Industrial Revolution*. Geneva: World Economic Forum._https://law.unimelb.edu.au/__data/assets/pdf_file/0005/3385454/Schwab-The_Fourth_Industrial_Revolution_Klaus_S.pdf

Schwartz, S. H. (1977). Normative influences on altruism. *Advances in Experimental Social Psychology, 10*, 221–279. DOI: 10.1016/S0065-2601(08)60358-5

Schwing, M., Reit, V., Selinka, S., Kuhn, M., & Meyer-Waarden, L. (2025). How Consumer Experience Shapes the Adoption of Autonomous Technologies: A Longitudinal Study With Shared Autonomous Vehicles. *Journal of Consumer Behaviour, 24*(3), 1133–1153. DOI: 10.1002/cb.2471

Seikh, M. R., & Chatterjee, P. (2025). Sustainable Strategies for electric Vehicle adoption: A confidence level-based interval-valued spherical fuzzy MEREC-VIKOR approach. *Information Sciences, 121814*. Advance online publication. DOI: 10.1016/j.ins.2024.121814

Shaheen, S. A., & Pan, A. (2024). Behavioral and sociodemographic impacts of carsharing. In *Handbook of Travel Behaviour* (pp. 339–363). Edward Elgar Publishing. DOI: 10.4337/9781839105746.00026

Shah, S., Khan, K., & Ahmad, N. (2021). AI-based predictive maintenance of autonomous vehicles for enhanced safety and reliability. *Sensors (Basel), 21*(2), 438. DOI: 10.3390/s21020438

Shamsuddoha, M., Kashem, M. A., & Nasir, T. (2024). A review on Transportation 5.0: Advancing sustainable mobility through intelligent technology and renewable energy. *Preprints*. DOI: 10.20944/preprints202411.2043.v1

Shanmugaraja, T., Ruba, M., Priya, V. S., Venkatesh, T., Supriya, M., & Nallamuthu, M. (2023, September). ECG noise elimination using an FPGA version of a modified error normalized LMS adaptive filter. In *AIP Conference Proceedings, 2764(1), 060033*. AIP Publishing LLC. DOI: 10.1063/5.0144367

Sharma, R. K., Saran, V. B., Kanaujia, S., & Gupta, S. (2024, May). Evaluation of mppt controller performance using fuzzy logic design with wireless sensor node. In *2024 International Conference on Smart Systems for applications in Electrical Sciences (ICSSES)* (pp. 1-6). IEEE. DOI: 10.1109/ICSSES62373.2024.10561365

Shiva, C. K., Vedik, B., & Manoharan, G. (2024). *Sustainable smart transportation*. IGI Global., DOI: 10.4018/979-8-3693-2373-1.ch017

Shladover, S. E. (2017). Connected and automated vehicle systems: Introduction and overview. *Journal of Intelligent Transport Systems*, *22*(3), 190–200. DOI: 10.1080/15472450.2017.1336053

Shokouhyar, S., Shokoohyar, S., & Safari, S. (2020). Research on the influence of after-sales service quality factors on customer satisfaction. *Journal of Retailing and Consumer Services*, *56*, 102139. DOI: 10.1016/j.jretconser.2020.102139

Shukla, R. D., Tripathi, R. K., & Gupta, S. (2010, November). Power electronics applications in wind energy conversion system: A review. In *2010 International Conference on Power, Control and Embedded Systems* (pp. 1-6). IEEE. DOI: 10.1109/ICPCES.2010.5698663

Sinha, R., Sinha, K. K., Patel, M., Gupta, S., & Priya, S. (2024, July). Detection of leukemia disease using convolutional neural network. In 2024 5th International Conference on Image Processing and Capsule Networks (ICIPCN) (pp. 451-456). IEEE. DOI: 10.1109/ICIPCN63822.2024.00080

Sinha, R., Kaur, N., Gupta, S., & Thakur, P. (2023, November). Diagnosis of Parkinson's Disease using Hybrid Ensemble Technique. In *2023 International Conference on Ambient Intelligence, Knowledge Informatics and Industrial Electronics (AIKIIE)* (pp. 1-5). IEEE. DOI: 10.1109/AIKIIE60097.2023.10390458

Smit, J., Kreutzer, S., Moeller, C., & Carlberg, M. (2016). *Industry 4.0*. Directorate-General for Internal Policies, European Parliament. https://op.europa.eu/en/publication-detail/-/publication/1b970736-9acb-11e6-868c-01aa75ed71a1

Statista. (2023). Global supply of EV batteries for recycling 2020-2040 [Dataset]. https://www-statista-com.eu1.proxy.openathens.net/statistics/1415407/ev-battery-recycling-supply-worldwide/

Steg, L., & Vlek, C. (2009). Encouraging pro-environmental behaviour: An integrative review and research agenda. *Journal of Environmental Psychology*, *29*(3), 309–317. DOI: 10.1016/j.jenvp.2008.10.004

Stern, P. C. (1999). Information, incentives, and pro environmental consumer behavior. *Journal of Consumer Policy*, *22*(4), 461–478. DOI: 10.1023/A:1006211709570

Sturgeon, T. J., Van Biesebroeck, J., & Gereffi, G. (2008). Value chains, networks and clusters: Reframing the global automotive industry. *Journal of Economic Geography*, *8*(3), 297–321. DOI: 10.1093/jeg/lbn007

Su, D., Gu, Y., Du, Q., Zhou, W., & Huang, Y. (2020). Factors affecting user satisfaction with new energy vehicles: A field survey in Shanghai and Nanjing. *Journal of Environmental Management*, *270*, 110857. DOI: 10.1016/j.jenvman.2020.110857 PMID: 32721306

Sukhodolov, Y. A. (2019). The notion, essence, and peculiarities of Industry 4.0 as a sphere of industry. In Popkova, E. G. (Ed.), *Warsaw Industry 4.0: Industrial revolution of the 21st century*. Springer. DOI: 10.1007/978-3-319-94310-7_1

Summers, C. A., Smith, R. W., & Reczek, R. W. (2016). An audience of one: Behaviorally targeted ads as implied social labels. *The Journal of Consumer Research*, *43*(1), 156–178. DOI: 10.1093/jcr/ucw012

Sun, N. (2023, July 31). New energy charging problem caused anxiety? Zhejiang program to help this “pile” small matter... ... *China News - Zhejiang*. https://www.zj.chinanews.com.cn/nzxzl/2023-07-31/detail-ihcrqyau6706271.shtml

Tanner, C. (1999). Constraints on environmental behaviour. *Journal of Environmental Psychology*, *19*(2), 145–157. DOI: 10.1006/jevp.1999.0121

Tansini, A., Pavlovic, J., & Fontaras, G. (2022). Quantifying the real-world CO2 emissions and energy consumption of modern plug-in hybrid vehicles. *Journal of Cleaner Production*, *362*, 132191. DOI: 10.1016/j.jclepro.2022.132191

Tata Elxsi. (2024). Transforming automotive with CASE: Connected, autonomous, shared & electric solutions. https://www.tataelxsi.com/industries/automotive/c-a-s-e

Teece, D. J. (2018). Business models and dynamic capabilities. *Long Range Planning*, *51*(1), 40–49. DOI: 10.1016/j.lrp.2017.06.007

Teng, J., & Guo, Q. (2020). AI-enabled dynamic pricing for ride-sharing to reduce traffic congestion and carbon emissions. *European Journal of Operational Research*, *287*(2), 652–663. DOI: 10.1016/j.ejor.2020.04.053

The Tech Outlook. (2023). Connected, autonomous, shared and electric (CASE) market is expected to hit a highest CAGR of 19.2%: Global trends, type, revenue, size, share, manufacturers, key drivers, future scope, supply and demand side analysis. Press release.

Thøgersen, J., & Ölander, F. (2003). Spillover of environment-friendly consumer behaviour. *Journal of Environmental Psychology*, *23*(3), 225–236. DOI: 10.1016/S0272-4944(03)00018-5

Tianjin Development and Reform Commission. (2023, March 2). *Municipal Development and Reform Commission promotes charging infrastructure construction Driving consumption of NEVs*. https://fzgg.tj.gov.cn/xxfb/xwxx/202303/t20230302_6126366.html

Tianjin Municipal Finance Bureau. (2016). *Municipal Finance Bureau Municipal Science and Technology Commission on the issuance of Tianjin to promote the application of new energy vehicles local subsidy management interim measures notice* (No. Tianjin Financial Construction I [2016] No. 40). https://cz.tj.gov.cn/zwgk_53713/zcwjx/zcwjst/202007/t20200719_2962749.html

Tianjin Statistics Bureau. (2023). *Tianjin National Economic and Social Development Statistics Bulletin 2022*. https://stats.tj.gov.cn/tjsj_52032/tjgb/202303/t20230317_6142668.html

Toyota Motor Corporation. (2023). Toyota Environmental Challenge 2050. https://global.toyota.

Toyota. (2005, March 22). *Tianjin FAW Toyota Motor Co., Ltd. Begins Production of the Crown in China | Toyota Motor Corporation Official Global Website*. https://global.toyota/en/detail/251090

Transforming our world: the 2030 Agenda for Sustainable Development | Department of Economic and Social Affairs. (n.d.). https://sdgs.un.org/2030agenda

Tyagi, M., Gupta, S., & Khan, M. I. (2024, May). Reduction of voltage instability in complex power networks through the utilization of D-STATCOM. In 2024 4th International Conference on Advance Computing and Innovative Technologies in Engineering (ICACITE) (pp. 1662-1666). IEEE. DOI: 10.1109/ICACITE60783.2024.10616856

Uddin, S. M. F., Sabir, L. B., Kirmani, M. D., Kautish, P., Roubaud, D., & Grebinevych, O. (2024). Driving change: Understanding consumers' reasons influencing electric vehicle adoption from the lens of behavioural reasoning theory. *Journal of Environmental Management, 369*, 122277. DOI: 10.1016/j.jenvman.2024.122277 PMID: 39222587

UNCTAD. (2023). Digital transformation and transport equity.

UNEP. (2024). *United Nations Environment Programme Emissions Gap Report 2024*. https://www.unep.org/resources/emissions-gap-report-2024

United Nations Industrial Development Organization (UNIDO). (2022). *Industrial development report 2022: The future of industrialization in a post-pandemic world*. https://www.unido.org/resources-publications-flagship-publications-industrial-development-report

United Nations. (2022). The sustainable development goals report 2022. https://unstats.un.org/sdgs

Van Vliet, O. P., Kruithof, T., Turkenburg, W. C., & Faaij, A. P. (2010). Techno-economic comparison of series hybrid, plug-in hybrid, fuel cell and regular cars. *Journal of Power Sources, 195*(19), 6570–6585. DOI: 10.1016/j.jpowsour.2010.04.077

Vasan, A., & Balamurugan, K. (2021). Autonomous vehicles and their impact on sustainable transportation. In Pal, S. K. (Ed.), *Intelligent Transportation Systems* (pp. 201–220). Springer. DOI: 10.1007/978-981-15-7798-9_10

Velaoras, K., Menegaki, A. N., Polyzos, S., & Gotzamani, K. (2025). The role of environmental certification in the hospitality industry: Assessing sustainability, consumer preferences, and the economic impact. *Sustainability (Basel), 17*(2), 650. DOI: 10.3390/su17020650

Vermesan, O., John, R., Pype, P., Daalderop, G., Kriegel, K., Mitic, G., Lorentz, V., Bahr, R., Sand, H. E., Bockrath, S., & Waldhör, S. (2021). Automotive intelligence embedded in electric connected autonomous and shared vehicles technology for sustainable green mobility. *Frontiers in Future Transportation, 2*, 688482. Advance online publication. DOI: 10.3389/ffutr.2021.688482

Verplanken, B., & Aarts, H. (1999). Habit, Attitude, and Planned Behaviour: Is Habit an Empty Construct or an Interesting Case of Goal-directed Automaticity? *European Review of Social Psychology, 10*(1), 101–134. DOI: 10.1080/14792779943000035

Verplanken, B., Aarts, H., Van Knippenberg, A., & Van Knippenberg, C. (1994). Attitude versus General Habit: Antecedents of Travel Mode choice. *Journal of Applied Social Psychology, 24*(4), 285–300. DOI: 10.1111/j.1559-1816.1994.tb00583.x

Verplanken, B., & Orbell, S. (2003). Reflections on Past Behavior: A Self-Report Index of Habit strength. *Journal of Applied Social Psychology*, *33*(6), 1313–1330. DOI: 10.1111/j.1559-1816.2003.tb01951.x

Wade, R. (1990). *Governing the Market: Economic Theory and the Role of Government in East Asian industrialization*. Princeton University Press., https://www.researchgate.net/publication/272933957_Governing_the_Market_Economic_Theory_and_the_Role_of_Government_in_East_Asian_Industrialization DOI: 10.1515/9780691187181

Waluya, A. I., Iqbal, M. A., & Indradewa, R. (2019). How product quality, brand image, and customer satisfaction affect the purchase decisions of Indonesian automotive customers. *International Journal of Services. Economics and Management*, *10*(2), 177–193.

Wan, X. (2022, January 20). Tianjin Consumers Association survey shows: New energy vehicle charging pile should also "charge charge." *China Consumers' Daily*. http://auto.ce.cn/auto/gundong/202201/20/t20220120_37272958.shtml

Wang, X. (2024). *The indispensable role of industrial policy: The case of the Chinese EV industry*. https://www.osservatorio-economie-emergenti-torino.it/emerging-economies/78-26-january-2024/443-the-indispensable-role-of-industrial-policy-the-case-of-the-chinese-ev-industry.html

Wang, K., Zhang, Q., & He, G. (2021). Sustainable mobility in smart cities: A review of AI-enabled solutions. *Sustainable Cities and Society*, *66*, 102657. DOI: 10.1016/j.scs.2021.102657

Wang, Y., Li, Q., & Sun, Z. (2018). Artificial intelligence in intelligent transportation systems: A review of applications, challenges, and opportunities. *IEEE Access : Practical Innovations, Open Solutions*, *6*, 62968–62985. DOI: 10.1109/ACCESS.2018.2877990

Westaby, J. D. (2005). Behavioral reasoning theory: Identifying new linkages underlying intentions and behavior. *Organizational Behavior and Human Decision Processes*, *98*(2), 97–120. DOI: 10.1016/j.obhdp.2005.07.003

Womack, J. P., Jones, D. T., & Roos, D. (1990). *The machine that changed the world: The story of lean production*. Rawson Associates.

World Bank Open Data. (2018). [Dataset]. https://datos.bancomundial.org/indicator/SP.URB.TOTL.IN.ZS?locations=ES

World Economic Forum. (2023). Circular cars initiative: Building sustainable mobility ecosystems. https://www.weforum.org

Xiong, R., Zhang, Y., Wang, J., He, H., Peng, S., & Pecht, M. (2018). Lithium-Ion battery health prognosis based on a real battery management system used in electric vehicles. *IEEE Transactions on Vehicular Technology, 68*(5), 4110–4121. DOI: 10.1109/TVT.2018.2864688

Xu, L., Mak, S., Proselkov, Y., & Brintrup, A. (2024). Towards autonomous supply chains: Definition, characteristics, conceptual framework, and autonomy levels. *Journal of Industrial Information Integration, 42*, 100698. Advance online publication. DOI: 10.1016/j.jii.2024.100698

Yilmaz, V., & Sürmelioğlu, Y. (2024), "Measuring automobile service quality with the European customer satisfaction index model (ECSI): the moderating effect of trust", *The TQM Journal*, Vol. ahead-of-print No. ahead-of-print. DOI: 10.1108/TQM-10-2023-0315

Young, W., Hwang, K., McDonald, S., & Oates, C. J. (2009). Sustainable consumption: Green consumer behaviour when purchasing products. *Sustainable Development (Bradford), 18*(1), 20–31. DOI: 10.1002/sd.394

Zhang, J., Zhang, C., & Zhang, Y. (2020). AI-driven logistics and supply chain management for sustainable transportation. *Transportation Research Part E, Logistics and Transportation Review, 138*, 101968. DOI: 10.1016/j.tre.2020.101968

Zhao, X., Li, X., Jiao, D., Mao, Y., Sun, J., & Liu, G. (2024). Policy incentives and electric vehicle adoption in China: From a perspective of policy mixes. *Transportation Research Part A, Policy and Practice, 190*(C), 104235. https://ideas.repec.org//a/eee/transa/v190y2024ics0965856424002830.html. DOI: 10.1016/j.tra.2024.104235

Zouari, G., & Abdelhedi, M. (2021). Customer satisfaction in the digital era: Evidence from Islamic banking. *Journal of Innovation and Entrepreneurship, 10*(1), 1–18. DOI: 10.1186/s13731-021-00151-x

Zou, Y., Li, M., Shi, M., & Yang, B. (2021). Predictive maintenance for electric vehicles using machine learning: A case study. *Energy, 224*, 120150. DOI: 10.1016/j.energy.2021.120150

Zuo, M. (2022, October 28). *Business Geography of China's NEV Cities |Tianjin: An Old Auto Industry City Waiting for a Turnaround*. https://m.21jingji.com/article/20221028/herald/daf738e8ea30ff6087980e3e0d8fe4cf.html

About the Contributors

Hugo Pérez-Moure is a PhD cum laude in Economic Analysis and Business Strategy from the University of Vigo. He has two Master's Degrees in East Asian Studies from the Universidad Autónoma de Madrid and in International Trade from the University of Vigo. Despite his young age, he has collaborated in multiple European projects in a wide range of business areas such as entrepreneurship or mobility. In addition, he has several publications in high impact JCR and Scopus journals such as Technology in Society, European Research on Management and Business Economics, Futures, Industrial Marketing Managemet, Suma de Negocios and Dirección y Organización. It is also a member of various international research networks such as AEDEM, RIICO and GERPISA. As a result of these networks and attendance at conferences, some of his research has won awards such as the ESIC award for Knowledge Transfer to Business and Society.

Jesús F. Lampón is Industrial Engineer and PhD in Business Administration and is currently Professor of Business Organization at the University of Vigo. He has been principal investigator in more than fifty contracts and research projects with different institutions and companies. His research has been published in more than 40 journals such as International Journal of Production Research, Journal of Business Research, Production Planning & Control, Papers in Regional Science, Regional Studies or Journal of Manufacturing Technology Management. He is editor of European Research on Management and Business Economics and SAGE Open, and member of the editorial board of Kybernetes, International Journal of Automotive Technology and Management and Regional Science Policy & Practice. He has extensive experience in the mobility industry and was responsible for different activities related to the Supply Chain of the automobile manufacturer Stellantis. He was Manager of the Program for the Promotion of Research and Business Innovation in the Consellería de Economía e Industria of the Xunta de Galicia (regional goverment)

***

Jorge Carrillo is Professor Emeritus of the Department of Social Studies at El Colegio de la Frontera Norte and a member of the National System of Researchers, Emeritus Level. He holds a PhD in Sociology from El Colegio de México. He has conducted research stays in Spain, France, Japan, and the United States. He is highly productive: author of 17 books; coordinator of 30 books; 177 book chapters; and 94 scientific articles in Spanish, English, German, Portuguese, Italian, French, Chinese, and Japanese. He has participated in 80 scientific research and consulting projects. He is a pioneer and widely recognized in Mexico and abroad for his studies on the Mexican export maquiladora industry. His current research interests are employment, innovation, and value chains in multinational corporations in Mexico.

Kashvi Chaturvedi, is studying B.Tech, specialisation in Artificial Intelligence and Data Science at Ajeenkya DY Patil University. She started solidifying her programming skills by completing various certification courses like Data Structures in C, Tableau functions and Operating Systems along with doing few projects using languages like Python, C, C++ and creating Web Page using HTML, CSS and Java Script. She had also attended an Arduino-Based Automation course at Dolphin Labs to enhance her software knowledge. There, she learned how to work with sensors, microcontrollers, and circuit design. She has also written multiple review papers on topics such as Geographic Information Systems, Comparative Analysis of BJT Configurations, Swiss Bank vs. European Central Bank: Data Analysis. She is also researching on and learning many topics such as Small Modular Reactors, Bibliometric Analysis, UNSDGs etc. Because of her interest in latest technologies, she also participated in Global Hyperloop Competition hosted by IIT Madras representing her college as team Vac-Vectoor where she learnt more about various subtopics from the mentors.

Susanta Das is a seasoned educator with a proven record of teaching and mentoring diverse students, working at multiple institutions, and providing service to the various initiatives of organizations. He received his Ph.D. and M.A. degrees from Western Michigan University (WMU), USA, and M.Sc. degree from Banaras Hindu University (BHU), India all in Physics. He continued his research as a Marie-Curie post-doctoral fellow at Stockholm University, Stockholm, Sweden on beam diagnostics for the DESIREE (Double ElectroStatic Ion Ring ExpEriment) in the project DITANET (Diagnostic Techniques for particle Accelerators – a Marie-Curie initial training NETwork), at the Indian Institute of Science Education and Research-Kolkata, India on high-pressure physics as a Project Scientist-B, and at the University of Electro-Communications, Tokyo, Japan on ion-surface interactions as a

Post-doctoral fellow. Throughout his career, Dr. Das worked and collaborated with many researchers, Ph.D., master, and visiting students from different countries. He further visited many countries to discuss research and an international conference participant (UK, Italy, Germany, Greece, Belgium, Bulgaria, Romania, Brazil etc.). He worked at several institutions before assuming his current role at Ajeenkya D Y Patil University. He received the Marie-Curie post-doctoral fellowship in Sweden, Science Academies' Summer Research Fellowship in India, Gowen Frostic Doctoral Fellowship, Department Graduate Research and Creative Scholar Award by WMU, and Leo R. Parpart Doctoral Fellowship by Dept. of Physics, WMU, among many others, throughout his academic journey. He co-authored several research articles in WoS/Scopus indexed international journals, conference proceedings, and scholarly book chapters and co-edited several books published by renowned international publishing houses (IGI-Global, Springer Nature, Nova Science Publishers, Wiley, Scrivener-Wiley, Notion Press). At present, Dr. Das continues to delve into cutting-edge fields, with a keen interest in nanotechnology, UN SDGs, and data science. His multifaceted contributions, spanning teaching, research, and administrative leadership, showcase a dedicated professional who is instrumental in advancing the vision and mission of the institutions he serves.

Haomin Fu is a PhD candidate in Automotive Science at Kyushu University, Japan. She holds a Master of Engineering in Automotive Science from Kyushu University and a Master of Arts in Linguistics from the University of Konstanz, Germany. Her research focuses on the downstream value chain of electric vehicles, with particular interest in sustainable mobility, digital transformation, and innovation strategies in the automotive industry.

Sandeep Gupta is presently working as an Associate Professor at Graphic Era (Deemed to be University), Dehradun (India). He has a total of 15 years of experience. His areas of interest in research are the application of artificial intelligence to power system control design, FACTS devices, renewable energy, power electronics, and stability of power systems with machine learning. He is a member of numerous professional organizations, including IE, IRED, IDES, IAENG, IAER, and the Society of Professional Engineers. He has been the author or co-author of more than 150 papers published in international journals and presented at different national and international conferences. He is currently guiding the research scholars under Machine Learning, Control & Power Systems, Wireless Technology, and the Internet of Things.

Mohammad Irfan is presently working as a Professor at School of Business, Chandigarh University, India. Dr. Irfan has done his Ph.D. from the Central

University of Haryana. He is MBA (Finance), M.Com (Account and Law), and MA (Economics). He has qualified UGC-JRF/SRF/NET in Management and Commerce. Dr. Irfan certified NSEs (NCFM) and BSEs certification. He has experience of more than sixteen years in the area of SAPM, Artificial Intelligence, Machine Learning, Blockchain, Cryptocurrency, Financial Engineering, Fintech, Green Finance, and Alternative Finance. He has to his credit more than 80+ Scopus-indexed articles, including The Journal of Economic Cooperation and Development (Q2), International Journal of Business Excellence (IJBEX), International Journal of Economics and Management (IJEM), Montenegrin Journal of Economics (Q2), Cogent Business & Management (Taylor & Francis) (Q2), Indian Journal of Finance (IJF), and Journal of Islamic Monetary Economics and Finance (JIMF). His citations reached 560+ along with 13 H-index. Dr. Irfan has published 11 books in Springer, IGI Global Publication (Scopus indexed).

Adriana Martínez holds a PhD in Social Studies from the Iztapalapa Autonomous Metropolitan University. She is a full-time professor at the Autonomous University of Mexico (UNAM): National School of Higher Studies, León Unit (ENES León). She is a member of the National System of Researchers, Level II.

Dhana S is a vibrant, enthusiastic professional with nearly 17 years of Academic, Research and Industry experience in institution approved by AICTE/UGC. Ample experience in academic, research, also curriculum design in Management, Tourism and Travel Management courses. Author for book "Multirater Feedback for Succession Planning". Proven researcher on various aspects of Management especially Human Resource, Marketing, Tourism and Hospitality Journals. She has published many articles in Scopus Q1,Q2 Journals, UGC Care List, WOS and reputed Journals. She is serving as Editorial Member and Reviewer for Elsevier (A*category) Reputed Journals. Attended Refresher Courses/STTP/FDP/Workshop Organized by MHRD, HRDC, UGC, AICTE, ATAL, etc., Also she Organized HR Conclave, International Conferences, Seminars, workshop, conferences, industrial visits, internship, OBT. FDP for Researcher and Faculty Members. Member in Board of Studies and various professional bodies like IAAC Society, ITHC. She also visited various countries like Singapore and Malaysia. Guiding doctoral Scholars and various administrative activities like NBA, AACSB, NAAC, IQAC, Community Social Service activities, organizing conference/seminar and workshops, she received notable awards like Young Faculty and Researcher award by NFED and Best Women Faculty award. Acted as a resource person for many FDP (Binary Accreditation for NAAC, AI Tool for Innovative Teaching) and Keynote Speaker for International Conferences. Courses Taught: Organizational Behavior, Strategic Management, Human Resource Management, Strategic Human Resource

Management, and Performance Management System. Talent Management, Business Law, HR Analytics, Entrepreneurship Development and Services Marketing. Research Interests: Organizational Behavior, HR Analytics, Entreprenuership, Sustainability, Tourism and Marketing.

Chandra Shekhar is a known academician with high level of knowledge & skill. He is Head of Department of an esteemed organization. His contribution towards any work is beyond any comparison. His performance is par excellence in every area of his expertise which includes Operations Management, Management Information System, Research Methodology & General Management. He is deeply involved in research and publications. He is a true mentor and guiding force of students & faculty members of department.

Devendra Kumar Tiwari is the Principal of United Institute of Management (FUGS), Prayagraj, India. He is an accomplished academic leader and researcher with wide-ranging interests spanning management, economics, behavioral sciences, and Indian knowledge systems. Dr. Tiwari has actively contributed to interdisciplinary research focusing on the integration of traditional wisdom with modern management and policy practices. His academic vision emphasizes innovation, ethical leadership, and holistic development through research, experiential learning, and institutional excellence.

Index

L

M

N

P

R

S

T

U

V

www.ingramcontent.com/pod-product-compliance
Lightning Source LLC
LaVergne TN
LVHW080855300926
849162LV00003B/429
* 9 7 9 8 3 3 7 3 4 4 9 5 9 *